T0292238

Did Extraterrestrials Bring Us to Intelligence on Our Planet?

A Scientist Speculates on the Sparse Information Available from Prehistory

John F. Caddy Ph.D.

Copyright © 2019 by John F. Caddy Ph.D.. 788792

Library of Congress Control Number: 2019902971

ISBN: Softcover   978-1-5434-9374-0
         Hardcover   978-1-5434-9375-7
         EBook   978-1-5434-9376-4

All rights reserved. No part of this book may be
reproduced or transmitted in any form or by any means,
electronic or mechanical, including photocopying, recording,
or by any information storage and retrieval system, without
permission in writing from the copyright owner.

Print information available on the last page

Rev. date: 08/12/2019

To order additional copies of this book, contact:
Xlibris
0800-056-3182
www.xlibrispublishing.co.uk
Orders@ Xlibrispublishing.co.uk

Contents

Did Extraterrestrials Bring Us to Intelligence on Our Planet?

Did Extraterrestrials Bring Us to Intelligence on Our Planet?

A Scientist Speculates on the Sparse Information Available from Prehistory

John F. Caddy

Introduction

I thought the most convincing place to begin this book was with the work of Zacharia Sitchin[1] on the Sumerians and their patrons the Anunnaki, while amplifying the wide range of sources he used. Thus, the structure of the book is reflected in the sequence of references at the end, which begins with the key analyses that provoked this study, and then lists more or less in sequence, the sources I discovered over the last year of research and writing.

Sitchin,[1] Knight and Butler,[2] and Hancock[3] are the authors whose works encouraged my interest in human prehistory, and as an ecologist, led me to search for other clues as to our origins. After an extensive literature search, I realized I did not agree with all of their conclusions, particularly with respect to their apparent exclusion of extraterrestrial influences on our history. It became obvious however, that we need an overall hypothesis as to the events that preceded what we call 'history'. Extending our mental time frame to make sense of human evolution beyond a field description from ancient sites requires immersion in a wide range of information sources. We need to rely on archaeology, geology, and even cosmology, in order to gain a broader perspective. The new ideas offered recently by human genetics also help pin down key events in our prehistory. How we put all this together mentally will vary from person to person, and throughout this book, I stress that what you are reading is *not* a dogmatic set of conclusions but one of several possible interpretations of a sparse network of facts. Only tentative conclusions on human prehistory are possible now, since information is scarce and interpretations uncertain, and I risk being accused of having an overactive imagination by presenting my own conclusions. Just the same, after reviewing what has been published on this subject, I believe that an overall hypothesis that presents a plausible picture of past events will be a useful basis for further discussion.

Why should we start with the work of Sitchin? His life's investigations are already public knowledge, however there seems a curtain of indifference, perhaps even fear, in accepting his hypothesis of extraterrestrial (ET) intervention early in human prehistory, and this is echoed by other key authors mentioned, who studiously avoid the subject of ET's, which until recently has been forbidden as a subject of serious discussion. There's also some concern expressed about the reactions of religious authorities and fanatics if a new interpretation of human development is offered. In fact, the development of religions in the Middle East millennia ago at least in part, could be why many important facts concerning our mental development on this planet have been hidden or remained below the surface. Recent information is only now beginning to emerge on other intelligences in the cosmos who have visited the Earth, and important persons or institutions apparently find this uncomfortable reading and would prefer to dispose of this subject as 'fake news'. After all, we live at a time when an excess of indiscriminate information/disinformation is available to everyone on the Internet, and this can rapidly give rise to apparently irrational conclusions.

It was comforting therefore, even when the text was almost finished, to receive a reference[95] from my daughter which concluded that a belief in nonhuman intelligence is increasing in unprecedented ways. Many contemporary technopreneurs are being inspired in their work by it. Mention was made of those people who believe in UFOs or extraterrestrials, characterized by Stephen Hawking as "cranks" or fringe dwellers. Despite that assertion, some of the world's brilliant Nobel Prize winning minds, among them the mathematician John Nash and the biochemist Kary Mullis, had experiences they perceive to be close encounters. The University of Oxford's Richard Dawkins, famous for his advocacy of Darwin's theory of evolution as well as his disbelief in God and religions, has suggested that human civilization may have been seeded by an alien civilization.

The radio broadcast of H.G. Wells' fantasy, 'The War of the Worlds' in 1938 reportedly led many people to flee their homes to escape the invaders from Mars [102]. This broadcast has been cited as the motive for governmental secrecy on the subject of visiting 'aliens'. In recent years however, Hollywood blockbuster films such as Star Wars have captured the public imagination, revealing a fairly widespread acceptance of the reality of cosmic life forms. Hence official secrecy on phenomena such as artificial structures observed on other planets of the solar system, has other motives that we can only speculate on.

My personal view as a scientist is that avoiding the investigation of potentially alarming phenomena can strongly bias our picture of reality, and that actions based on ignorance are to say the least, dangerous. One common practice I have encountered in preparing this text is the tendency of many experts to state emphatically that some discovery of a predecessor (or contemporary) expert was incorrect or misinterpreted. As a non-specialist, my approach is inevitably less judgmental—I evaluate new information in terms of its possible impact on the overall emerging hypothesis. The final result is of course now available for modification by experts, hopefully those with a broad experience of different disciplines!

Do we ignore the conclusions of Zaccharia Sitchin on the origins of our species because some suggest that they are based on false research, or is it because we don't want to accept his opinion that in human prehistory, we were modified by a more advanced species? I suppose a common viewpoint is that we evolved naturally by Darwinian processes into a relatively uncommon species of migratory hunter-gatherer, and then somehow accelerated ourselves into a high tech civilization? That we were modified and civilized by a species at a higher level of culture than ourselves to become one of the most abundant mammalian species on the planet, may seem a non-essential addition to standard Darwinism. However, reasons emerged in the course of this study that suggested that assistance from elsewhere was arriving to speed up our evolution, and help overcome certain inhospitable issues we face from our planet of origin. Would accepting that help may be forthcoming lead us to take less pride in our own achievements and hence be unacceptable? The approach I have taken to this conundrum is to imagine that whatever conclusion the available information suggests when it is put together, as long as it is coherent, it forms a useful (and hopefully entertaining!) story that can be elaborated upon or contradicted by new information. The specialists will inevitably question my sources and conclusions, but I suggest that we might all benefit from further investigations of our origins by open-minded non-specialists.

What Dr Sitchin[1] told us years ago still hasn't percolated through to society, and this suggests that strong opposition exists to his earth-shaking conclusions. The logic that the cosmos has already generated civilizations much older and more sophisticated than ours can no longer can be contradicted by scepticism. Hence, their existence is an established axiom for this study—as well as a conclusion based on some limited personal experience. It seems improbable that this complex and beautiful planet passed through several billion years during which it was ignored by extraterrestrial intelligences whose evolution long preceding ours! The hypothesis that we were taken under the wing of one or several ancient and civilized cosmic species—and were obliged to pass through an 'apprenticeship' to become more intelligent and civilized beings—may not be a comforting conclusion to some, but follows from some convincing indications. As you will learn later, our early apprenticeship left us with some less than ideal characteristics we should try to correct. Nonetheless, a cosmic view of life now seems appropriate now that we are seriously involved in the exploration of space.

Dr Sitchin's competence as a translator of Sumerian texts came from working in that field for many years, and I give him credit as a courageous expert who looked outside his specialization for interpretations. As an ecologist, I've looked for further evidence outside the basic information sources he relied on: the Sumerian clay tablets and other sources of information on early civilizations. There is enough data now available to support the idea that evolution took us part of the way to humanity, but that extraterrestrial influences on life's evolution, and on our species, have been significant during our most recent voyage towards civilization. We need to better understand the sudden change from isolated tribes of a rather uncommon species of hunter-gatherer, to inhabitants of cities of 10 million people or more.

Our Planet as an Incubator of Life

At the time of the Apollo adventure on the moon one of the most thrilling and daunting images we saw was the first vision of our planet from space—a blue sphere wrapped with clouds which marked the fragile boundary: a thin film of air covering our planet which separates us from the vacuum of space extending out to infinity. As we grow more accustomed to the idea that we depend on the only breathable atmosphere in the solar system, we should become more aware of our collective responsibility for conserving it. This should also affect how we view our evolutionary context and our potential relationship to other intelligent species in the universe who are capable of moving freely through the vacuum. In these circumstances, we need to slowly work towards a cosmic perspective, recognizing the reality that because our knowledge of the cosmos is limited, we need an imaginative interpretation of the limited facts available. With a sober state of mind, this approach can lead to useful hypotheses; for example, those expressed in the books by Knight and Butler.[2,3] These suggest that a superior civilization existed before the Great Flood, a conclusion already arrived at by Graham Hancock.[4] Knight and Butler came to this conclusion after realizing the ancient, sophisticated, and geo-physically integrated nature of the everyday system of physical measures we use, and how it would have been effectively impossible for primitive mankind to arrive at these essential measurement tools without receiving advanced information from those we might call 'tutors'.

Knight and Butler as authors started with a long-forgotten unit of length, the megalithic yard, derived by Alexander Thom from measurements of megalithic monuments. They demonstrated that our everyday measures of length, weight, and time are connected to this forgotten measure of length. This, in turn, is tied to the dimensions of the solar system. The hypothesis of Knight and Butler is that our 'common or garden' units of pounds, pints, minutes, and other humble measures were derived millennia ago from geo-statistical measurements of our planet, moon, and solar system by a space-going intelligence. An early advanced source also left behind mental relicts in the form of mythologies to help societies reconstruct themselves after the worldwide Great Flood. Graham Hancock also suggested that these advanced sources, or their survivors, offered more concrete assistance during their visits to regional centres after the Flood, to help restore human civilizations back to their former levels of sophistication.

In their second book, Knight and Butler discussed the much earlier 'construction' of the Moon, which, in their opinion, was undertaken by a higher order of extraterrestrials to encourage the evolution of intelligent life on this planet. This conclusion is not widely accepted and is not essential to my hypothesis; however, controversy as to the nature and origin of the Moon persists and cannot be resolved here. Nonetheless, it is clear that the moon was vital in evolution, given that tidal forces affect environments and life cycles and make chemical components accessible to life forms.

There is now a growing understanding that many forms of intelligent life may have preceded us in the universe, and that they had sufficient time and valid reasons to promote the evolution of their intelligent successors. Of course, we face the common conviction that life on Earth is a unique phenomenon. However, what is called directed panspermia,[5] that is, the deliberate introduction of DNA or organisms containing it onto this planet, has become the likely alternative to what would have been a slow and tortuous evolution if all stages of it had to be completed here by Darwinian processes. In either eventuality, the idea of Knight and Butler was that our planet was modified to become an 'incubator' to breed intelligent animals. Given the aeons of time available, one might add that the desired intelligent products of 'Cosmic horticulture' might have included more than smart mammals. The assumption that it took 3.8 billion years since life began here to produce an intelligent species seems frankly pessimistic, given the relatively short period of 1.5 million years primates took to evolve into quasi-civilized humans. In my view, the rapidity of this evolution implies the existence of advanced life forms from elsewhere, who for whatever motive, decided on this strategy. This last comment gives me the opportunity to state that in my opinion, the entities who may have been responsible for geo-transforming our planet, are not necessarily those who implemented panspermia, and almost certainly not those who much later, brought us to civilization.

Sitchin's hypothesis that genetic upgrading was the mechanism for drastically shortening our evolution to the city-dwellers of our global technological society within an interval of not much more than ten to fifteen thousand years. He revealed the genetic procedures which led to this rapid evolution of intelligence, after which we were taught the new techniques and social structures needed to produce, store, elaborate, and transport the large quantities of food and other raw materials large sedentary communities required.[6] At this point, I believe that a key idea that will help us towards a cosmic consciousness can be derived from the concept of ontogeny, which is the ensemble of successive stages in the life history of an organism. I use the term 'evolutionary ontogeny' to describe the many stages that preceded the final species an organism arrived at before it's eventual extinction. We seem to assume that humanity has arrived at the end point of our 'evolutionary ontogeny', but I am quite sure that bigger changes will come after some of us leave the Earth behind.

A similar hypothesis suggesting an even earlier intervention on human prehistory by another extraterrestrial species was proposed by Fenton,[6] which she supposed was also aimed at improving our mental capacity. It should be borne in mind that as many as nineteen extraterrestrial species have recently been identified visiting our planet.[12] Therefore, what Sitchin or Fenton proposed may form only a fragment of the unrecorded prehistorical encounters between our species and extraterrestrials. We remain largely unaware of what these influences were, but a recent spate of reports in the media and Internet have focused on past hypothetical off-planet influences on our religious and spiritual beliefs, suggesting that, in some cases, these may have originated from visits by beneficial extraterrestrials.

The key catastrophe that effectively wiped out most civilizations and their historical records that hypothetically existed 12,000–13,000 years ago, was the global mega-flood, and after that is when documented history begins. Some experts believe this was only one of repeated geological/cosmological traumas experienced by our planet at long intervals, but this last event seems to have been an evolutionary 'watershed' dividing the unknown early civilizations from the better-known civilizations present after the great flood. Sitchin's work, and that of others who followed, inspired me to support an existing framework suggested by others, namely that the human social evolution to urban dwellers was assisted by another intelligent species which was at a much later stage of social development than we were. This framework invites other suggestions for alternative prehistorical frameworks—if they can be synthesized. Concentrating on developing an overall framework, even if it contains numerous gaps in knowledge, rather than simply accumulating random pieces of specialized information may be the only way we can make progress in understanding key events in our past as a species. This is especially the case if the authorities in relevant fields have strong motives for supporting current more limited historical interpretations.

Apart from what might simply be called 'fear of the unknown' as a reason the public may not wish to admit to the existence of giants and off-planet beings, there is also a growing and rational fear of the 'fake phenomenon'—that is, a widespread aversion to deliberate attempts to hide new information sources and block starting assumptions which postulate ET activity on Earh. Of course many issues brought up on the web when subject to even superficial inspection seem wildly exaggerated and transparently improbable, and may well be so. There is also the fact that governments and technological industries maintain secrecy on issues related to ETs and space technologies, since these have significant military implications, and this may have blocked information searches involving ET's. The reason for secrecy is especially evident since at least some among the recent burst of discoveries in aerospace and electronics, seem to have come from the reverse engineering of crashed extraterrestrial vehicles.

Our current technological civilization is now on the verge of routine spaceflight to other planets, and this sudden increase in our technical capability may be essential to the long-term survival and further evolution of our species. The diagram on the cover of this book is intended to suggest that a changeover occurred from the Darwinian evolution of the products of panspermic DNA dispersed by ETs mega-years ago, to a more directed evolution with the goal of making us space-going under the guidance of a still further evolved species. A process that we may call Syntropy appears to be at play here, implying that we, and events, are being drawn onwards by a mysterious influence coming from the future.[80,84] This seems to have taken over from Darwin's model in influencing our likely evolution from the planet's surface to the immensity of outer space.

It is useful to bear in mind that there may have been other 'prehistoric' civilizations of humans or other intelligent species living on this planet prior to the last Ice Age, and it seems feasible to suppose that they followed, and perhaps completed, a similar trajectory to the one we are now on? This may even be likely given the vast time interval that passed before our appearance. Another possibility to consider is that other species that did not originally evolve here may have come to inhabit this complex and productive planet. What then does 'history' tell us about these possibilities? Virtually nothing, since the above suppositions are taboo in academic circles. Underlying the hyper-imaginative historical hypothesis shown in figure 15 (which anticipates much that you need to read first in this book!) is the vast time span from the origin of life on Earth until the last Ice Age. This figure is presented with one main objective which is *not* historical accuracy, but an attempt to show that the period when historical information on our own history is available, is trivially short. Many events could have happened in the distant past, a few of which are imagined on this figure based on the few scraps of information we have on our prehistory. Thus, we need at least two time scales to represent past events—geological and historical. 'History' (i.e., past events involving identified persons or societies of our species) does not extend much further back than the Great Flood.

In a paper entitled 'The Silurian Hypothesis', scientists at NASA and the University of Rochester[8] looked for evidence that our civilization was preceded by an advanced civilization of intelligent reptiles. 'Do we really know that we were the first technological species on Earth?' asked Adam Frank, a co-author of that paper: 'We've had an industrial society for only about 300 years, but there's been complex life on land for nearly 400 million years.' Likewise, a future civilization arising millions of years from now might find few traces of our human civilization (except perhaps a few CDs in the stratum of non-biodegradable plastic we will have left behind)!

A repetition of global temperature trends in successive epochs

I'd like to draw attention to past quasi-cyclic global temperature trends in the long term history of our planet as revealed by analysis of long term ice cores taken by drilling down from the surface in Antarctica [93] (see fig 1). Expert analysis revealed annual strata whose characteristics allowed experts to deduce the state of the environment in past epochs. The apparent periodicity in global temperatures over the long term emerges from the following graph of temperature time series deduced from the characteristics of ice samples in the cores. These extended down for thousands of metres, cutting through ice layers that accumulated as seasonal snow fall in the distant past. An informal way of illustrating the apparently cyclical nature of climate change over the very long term, before our species came on the scene, was illustrated by dividing the whole time series of some half million years or so, into what I have called 'epochs'. Each epoch was envisaged as starting just prior to the high temperature spikes which show up at regular intervals throughout the time series, after which temperatures oscillate but slowly decline, eventually arriving at cold periods or Ice Ages. Each epoch derived in this fashion lasted approximately 100,000 years, and the duration of these epochs seems to have slightly increased over this time period (see table 1).

I manually superimposed the trends in mean temperature for epochs A- D, starting at the lowest temperature period in each case, and marking the points where the trend changed direction in each epoch. The superimposed points for epochs A-D were painted blue in Fig 1 and used as a reference to judge events in the most recent period. Even though this is a 'rule of thumb' approach without statistical analysis, this procedure adequately illustrates that the temperature sequence in earlier epochs showed closely similar trends in each epoch, and were presumably determined by cosmic factors such as the changing inclination of the Earth's axis and/or changes in the sun's level of radiation, that we don't need to analyse further here. We are using the blue surface in the last square of Fig 1 to essentially represent the average temperature trend in earlier epochs, when existing evidence suggests that an industrial human civilization and its activities which affect the climate, were absent. In comparison, the temperature trend so far during the last epoch is a clear indication that anthropogenic warming has so far increased temperatures steadily above the range shown for the previous epochs. However if the declining temperature trend a-b for earlier eons is followed later on in our own epoch, in the long run this should help moderate warming impacts if we can survive the immediate heat pulse we have caused!

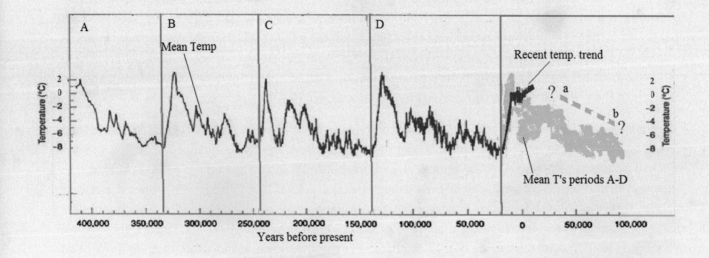

Fig 1. A long-term temperature record derived from analysis of ice cores in Antarctica [93].
The division of the temperature data series into epochs A, B, C and D, beginning at the low
point just before the sharp peaks in temperature, shows similar trends in each epoch and
their combined performance is shown by the blue surface in the most recent period. The
dotted orange line a-b shows how a downward trend in global temperatures is eventually
likely over the very long term even if the causes of overheating of the climate due to human
activities are not adequately tackled and temperatures remain higher than for earlier epochs.

Table 1. Approximate durations of epochs in the above figure (dates BC).

EPOCH	FROM	TO	DURATION
A	420,000	335,000	85,000
B	335,000	245,000	90,000
C	245,000	140,000	105,000
D	140,000	20,000	120,000

Catastrophes and Climate Change in History

The possibility of detecting past civilizations on Earth before 1 million years ago may be slim given the extreme dynamism of our planetary surface. In addition to global temperatures which seem likely to fluctuate in a broadly similar way to those shown for the Antarctic, other catastrophic events such as continental drift, earthquakes, landslides, wind and water erosion, sediment deposition, oxidation of organic traces by the atmosphere, volcanic eruptions, and meteor bombardments will all have erased most traces of previous intelligence or earlier civilizations that may have existed here thousands or millions of years ago. Regular planetary disasters seem characteristic of our planet, so that few if any ancient artefacts and constructions would be left a million years after a global civilization existed, and over even shorter periods given the past durations of moderate climate shown in Fig 1.

The surface of our planet is also regularly renewed by the subduction of its surface layers in the course of continental drift. The combination of all these factors means that the history of intelligence on our planet is hard to trace back beyond the last Ice Age some 13,000 years ago, except for a few exceptionally well-preserved relics. Even 'The King List', that remarkable chronology of Sumerian and pre-Sumerian kings[1] recorded on clay tablets extending back over 224,000 years (with ten kings before the Flood!) may not be a reliable chronicle if the individual lifespans of early rulers have been exaggerated. Nonetheless, for issues relating to the Anunnaki and other possible 'sponsors' of our move to civilization, it would be better to avoid excessive scepticism on matters beyond our human experience[8] and examine the available information unimpeded by the idea that we are experts on the matter.

One possibility that has been raised is that the evolution of a much earlier (reptilian?) intelligence occurred so long ago that it did not leave an identifiable fossil record, but we may need an explanation for their apparent absence from 'our' planet nowadays. As a sub-hypothesis (although it may initially appear absurd), consider Velociraptor; a medium-sized predatory dinosaur whose reconstruction from fossils we all saw in the film *Jurassic Park*. As a 'pre-intelligent' species it lived during the Late Cretaceous about 80–85 million years past—that is, before dinosaurs became extinct after a mega-meteor 10–15 km across landed in the Gulf of Mexico 66 million years ago. Arriving at a technologically sophisticated civilization within the 15 million years left them before planetary disaster occurred was not impossible, given that these medium-sized predators hunted in packs and were equipped with behavioural and morphological features which, in the more than adequate time available, could have given rise to an intelligent species. This species, or others like them, were social and bipedal, with clawed 'hands' potentially modifiable by evolution for manipulating objects. Who could deny that intellectual and technical sophistication was a potential for their future? The fact that their descendants, whom we now call winged dinosaurs (e.g., parrots and ravens), are intelligent does not prevent their distant un-winged ancestors millions of years ago from having evolved from a similar condition! It is easier to explain, for example, that the ancient manufactured artefacts occasionally found in coal mines from the Jurassic Period are the products of an industrial age of intelligent dinosaurs rather than to automatically assume that humans existed millions of years ago,[48] and have been in evolutionary stasis since then? The huge time intervals then available certainly allowed for other possibilities than a human species frozen in a million-year stasis, or perhaps equally improbable, colonies of a humanoid species transplanted back in time by an advanced species with expertise in time travel?

In contrast, consider[7] that the evolution of *H. australopithecus* to *H. sapiens* took only 1.5 million years of accelerated development! One reason for our more recent super-fast evolution to the modern sub-species H. sapiens sapiens, other than by natural selection alone, is a subject to be discussed later. If a superior species organized our evolution and is still active in the universe many millennia later, perhaps we should also wonder at their attitude to the collision of large astral bodies with the earth? Such an event would compromise any carefully organized evolutionary experiment which was being performed on us by advanced extraterrestrials, as Knight and Butler postulated. Protecting our planet from mega-disasters of this type would be child's play for a superior entity, and one may even ask why the extinction of the dinosaurs was allowed to occur 50 million years ago. Perhaps the survival and future evolution of mammals was a higher priority? Or perhaps intelligent reptiles had already left the planet by the time the Cretaceous-Tertiary extinction event finished the reign of the dinosaurs.

An example is given later on of how bombardment by meteorites from space could be a relatively simple and successful form of aggression by a space-borne species. We on the gravity well of our planet's surface would have difficulty countering such an attack. That human civilizations have not been overtly colonized by ETs since the time of the Anunnaki, suggests that the alarmism often expressed on this possibility does not reflect the true intentions of ET. Possibly ET may be expressing protective instincts, as revealed by the 'interventions' of UFOs to de-activate IBMs about to leave their launch pads?

The peaceful if intellectually challenging patterns left in fields of grain and elsewhere by ETs are also hard to consider acts of aggression. That natural disasters could be used by the 'Almighty' or his ET representatives, to terminate the existence of intelligent but undesirable inhabitants of the Earth, is hinted at in the story of Noah's Ark. This seems to be a myth reflecting a real event that occurred a thousand years or so before the Old Testament was written. In practical terms, the later biblical testimony refers to the mega-flooding that occurred at the end of the last Ice Age, which is recorded in other ancient accounts and traditions on a worldwide basis. More specifically, Sumerian writings[1] discussed the carelessness of the Anunnaki with respect to almost all of their human servants who were *not* embarked on the Ark (i.e., the misbehaving majority of genetically enhanced *Homo s. sapiens* they left to die in the Flood).

At this point, I should mention C. H. Hapgood, who wrote the book *The Earth's Shifting Crust* with James H. Campbell.[33] They are the best-known advocates of the controversial claim that a rapid and geologically recent pole shift led to sudden displacements of the earth's crust (the lithosphere). Some specialists refute this mechanism, just as contemporaries of its discoverer, Alfred Wegener, laughed at his hypothesis of continental drift. And just as Darwin's ideas on evolution scandalized the religious authorities at the time. Darwin's early ideas on evolution tended to follow those of the influential geologist Charles Lyell, who believed that evolution happens gradually. Nonetheless, with further research, catastrophic rather than gradual change is becoming recognized as a feature of the Earth's behaviour, and a similar mechanism may have resulted in catastrophic consequences at the close of the last Ice Age that are difficult to explain without either a sudden and extensive movement of the earth's crust, or of the Earth's axis of rotation. For example, an ancient chart referred to as the Piri Reis Map, supposedly dating from the fifth century, appeared on close inspection to be based on much older material, evidently coming from a technically sophisticated, ancient, and geographically well-informed source.[4] This map shows a large southern continent with rivers, ports, and trees (i.e., a land with a moderate climate) and with a coastline coinciding closely with the recent results of sonar mapping through the ice of the geological periphery of the continent we now call Antarctica. Other supporting evidence for a sharp shift in the earth's crust at the end of the last Ice Age is the presence of mammoth carcasses in Siberia, which dates from a sudden freeze that occurred there during the eleventh millennium BC following a rapid latitudinal shift, with all the chaos that this would have involved. In the 1900s, their meat was found to be still edible, but it had been torn apart, flash frozen and buried in mud with fresh grass still in their mouths. The close occurrence of a loss of the North American megafauna could be a parallel indicator of the rapid climate change that occurred at that time.

Such observations imply that organisms that were previously living in a moderate-cold environment climatically warmer than their current location, were shifted suddenly into full Arctic or Antarctic conditions. Presumably, the ice fields around the Antarctic continent were suddenly displaced in a massive northward travelling tsunami that led to worldwide flooding at around this date.

Recent satellite photos on the shore of the Ross Sea which have been exposed by the Antarctic's currently melting ice sheets, show large-scale images of humanoid faces inscribed in the terrain[34]. These will probably be used to support the long-held theory that the Antarctic was the site of Atlantis, but this was not necessarily a European or human phenomenon, even though Atlantis was first described by the ancient Greeks. Recent satellite photos of what appeared to be large-scale portraits on the rocky surface exposed by the melting of the snow and ice cover were accompanied by what have been identified as proto-Sumerian inscriptions, as well as evidence of visits by the early Tamils. These inscriptions both suggest that an advanced civilization was located there and had international (trade?) contacts with its inhabitants before a shift in the continental plates led to extensive ice formation rapidly annihilating their homeland. Egyptian visits to the southern hemisphere have also been identified by hieroglyphic inscriptions located in Australia. Are we seeing the first evidence of ancient exploration and trade in the southern hemisphere before the last Ice Age? Could it be that proto-Antarctica was a final Anunnaki 'home away from home', (as they called their first base in Mesopotamia), in what was originally a cold-temperate 'proto-Antarctic' continent lying to the north of the South Pole? Also, perhaps the climate catastrophe that enveloped this southern continent was what persuaded the survivors from this disaster, if there were any, to leave Earth for their more stable home planet.

Ancient Pre-Humans

A recent lecture on YouTube by archaeologists and anatomists[14] from Peru was illustrated by X-rays of several ancient embalmed bodies found there, dating back a millennium or so. These included the hominid forms discussed later in this article as 'Cone-Heads', but convincing X-rays were also shown of a small bipedal 'hominid' found near Nazca.[46] The embalmed remains of this last mentioned 'person', a hominid with three fingers, may have emerged as a reptilian, given that the X-rays show what appear to be several large eggs in its abdomen. No further information on this or similar specimens is available, so I cannot comment further, but the findings seem to support the hypothesis that a reptilian hominid was once at home on this planet.

According to Zecharia Sitchin, the Anunnaki (in Sumerian, 'those who from heaven to Earth came') were an extraterrestrial species who arrived on Earth well before the last Ice Age and perhaps not long after the one that preceded it. They appeared to be mammalian in form and are reported to have reigned for thousands of years in Mesopotamia. They mined gold in the Arabian Sea and later in Southern Africa and elsewhere,[1,10] using our forefathers for manpower after modifying them from pre-hominids[1]. Although Mesopotamia was the original point of arrival for the Anunnaki, the pre-humans they used as a labour force came from Africa, judging from the account of Michael Tellinger.[11] Most of the Anunnaki are believed to have left the Earth after the Flood. However, what may be their hybrids with humans have been identified in ancient archaeological excavations, and it seems reasonable to hypothesize that they later provided the technical expertise underlying Megalithic civilizations around the world.

The polygonal form of cyclopean masonry typical of Megalithic structures confuses modern experts, not only because of the unknown methods of shaping and transporting huge blocks of stone, but because it is unclear why they had to employ such an apparently labour-intensive approach to construction, instead of using more easily programmable rectangular building blocks? If one associates the remarkable stability of megalithic structures over the centuries, often in earthquake zones, in comparison with buildings constructed with small rectangular blocks of stone, and if one also accounts for the much longer life expectancy of the Anunnaki, the answer becomes obvious. Who would want to live in a house more likely to collapse due to an earthquake during their extremely long lifespans?

In an account of their mental communications through channelling with extraterrestrials, Scott Jones and Smith[12] tested a standard questionnaire on nineteen extraterrestrial species, one of them being an Anunnaki. When asked in this 'cosmic communication' if other extraterrestrials had criticized their species for using early humans as slaves, their Anunnaki correspondent replied that they did not rule over planets, although they could be accused of paternalism: 'Humans are our children; we worked together.' According to this witness, there are still a few of his species on Earth in apparently human form but not in dominant positions in society. This 'discussion' also suggested that an unspecified number of Anunnaki have since left their home planet for other locations in the galaxy. It should not surprise anyone that I cite studies such as this one, which used extrasensory perception and later the obviously related paranormal skill of 'remote viewing'[74] to investigate events distant in time and space.

It seems certain that we will eventually have access to such superior methodologies not yet in common use nowadays as our scientific knowledge and perceptive capability is vastly increased in scope through contact with extraterrestrials. Their evolution and knowledge base preceded ours by thousands of years or more, and a common theme on the internet explains the dramatic proliferation of microprocessors and other new technologies in the market place as a result of reverse engineering of alien technologies encountered in investigating flying saucer mishaps.

On the question of the sensory and intellectual capabilities of extraterrestrials such as telepathy, telekinesis and access to other dimensions, it may even be that some of these were components of our original sensory capabilities that we unwittingly renounced when we were modified by ETs, and later when we began to focus most of our intellectual energies on materialism. Elsewhere,[58] I suggested that our correspondence with extraterrestrial visitors might improve if we could relearn some of these paranormal capabilities.

Random and Directed Panspermia

Random panspermia (or the origin of life elsewhere) has gained credibility recently. This theory supposes that meteorites hitting Mars displaced biological material from this earlier site of evolution into space and brought DNA from Mars to our planet. This notice was greeted with enthusiasm by members of the public, who cried out, 'We are all Martians now!' Panspermia now seems difficult to dismiss and is not impossible in a galaxy that probably has many intelligent species, some of which must have evolved long before Homo sapiens, which according to the earliest skeletal remains found in Morocco dates back to 250,000-350,000 years ago. As such, we are defined as: 'Bipedal primates having language and the ability to make and use complex tools, and a brain volume of at least 1400 cc'. Homo sapiens sapiens is a very recent sub-species, this terminology often being used, not to imply a morphological distinction with our predecessor species, but referring to the intellectually-based pursuits modern man is now capable of in urban civilizations world-wide. Such urban dwellers are considered here to have originated through the intervention of the Anunnaki.

One characteristic of life on this planet is the way that plants and animals share similar genes. If DNA had arisen here by slow mixing of different original components of genetic material, perhaps, as one argument holds, there would be much more divergence in the DNA-basis of terrestrial organisms. Another perhaps better argument for panspermia is the fact that among the essential basic components for life here are elements such as molybdenum, which is relatively rare on Earth, while common elements here—such as lithium, chromium, and nickel—are not much used by living organisms. This might imply that evolution began on other planetary bodies where the essential elements for evolution were more abundant than here.

Francis Crick,[5] along with James Watson, is noted for discovering the structure of the DNA molecule. In 1953, he co-authored (with H. Orgel) the book *Life Itself*, in which they asserted that there is no possible way that the complex DNA molecule had time to evolve on Earth by natural processes from simple molecules. Therefore, they argue it must have come, already synthesized, from somewhere else. Some theoretical considerations that support this point of view were put forward by geneticists Richard Gordon and Alexei Sharov,[22] who deduced that if the evolution of life followed Moore's law, then life must predate the existence of planet Earth. They started with the idea of genetic complexity doubling every 376 million years. Working backwards, this would mean that life first arose almost 10 billion years ago. This would predate the creation of the earth itself, which was formed some 4.5 billion years ago. Assuming that Moore's law, derived from the evolution of computer chip complexity also roughly applies to biological complexity, life must have begun somewhere else and migrated, or was brought here (see Yerka[26] for a short summary of this argument).

Mixing Genotypes to Produce Homo Sapiens?

Recent DNA analysis is only now beginning to illustrate the complexities of human evolution. Ironically though, DNA is one of the most durable structures in nature; it is passed on from generation to generation with minor changes or 'mutations' in the genotype[15] and mitochondrial DNA. However, only a small proportion of the DNA spirals in human chromosomes are dedicated to genes; the rest are probably misleadingly called 'junk', or non-coding, DNA. This makes up the rest of the spirals, and it is currently difficult to interpret its function or history.[45] However, some scientists believe that this 'inactive' part of the complex molecule is not necessarily just a result of evolution, but they speculate it may contain a message from life's original designers[18] to the first life form who can find a way of reading it. Others suggest that the changes in the human genome Sitchin ascribed to the work of Enki (see later) were only intended to add a small proportion of the Anunnaki genes to ours, and that the 'junk DNA' may show where other Anunnaki (and pre-human?) genetic information was cancelled from the modified human DNA—or is this the space left after our original instinctive behavioural and sensory adaptations were removed to make us more peaceful and efficient workers? One might suppose that one objective of the Anunnaki, as genetic experts, was to erase genetic components that promote aggressive behaviour, as well as those sensibilities needed in the wild, and to improve our ability to obey orders and manipulate tools. These processes are broadly similar to what is referred to as 'domestication', a procedure we ourselves have used to selectively breed dependent species and pets into a more docile state.

Zeng et al.[72] noted that about 7,000 years ago, there was a dramatic collapse in the genetic diversity of the Y chromosomes of men, and that effectively there was only one man left to mate with each of the seventeen available women. They argued that the origins of the collapse (referred to as the 'Neolithic Y-chromosome bottleneck'), was the consequence of continuous wars between patrilineal clans descended from a very limited number of male ancestors in each tribe. The researchers hypothesized that if wars repeatedly wiped out the male members of entire clans over time, they would also wipe out many male lineages and restrict the variability of their unique Y chromosomes. A possible example of this mechanism was the Yamnaya/Beaker People invasion of Europe (described later). An alternative hypothesis comes to mind, however, which suggests a way of accounting for the current low diversity of the male genome in light of the roughly contemporaneous events discussed here. If, for example, our ancestors were conceived by artificial insemination using sperm from only a few male donors, could not this have also resulted in a reduction in the diversity of the Y-chromosome?

Neurobiologist Pierre Vanderhaegen[20] noted that we are separated from chimpanzees by 6 million years, but we carry within us only 200 different genes from the chimp genome out of the more than 20,000 genes we share in common. Of the new genes, three of them—NOTCH2NLA, NOTCH2NLB, and NOTCH2NLC—control brain tissue growth in the foetus. They act together to slow down the transformation of stem cells into neurons, and by slowing down this process over four months during our longer childhood than experienced by chimpanzees, this paradoxically results in more neurons and a larger brain. Further research by other neurobiologists shows that the sophistication of the human brain is not simply the result of steady evolution. According to new research (see Alok Jha[19]), human brain evolution developed extraordinarily fast: a speedy evolution unique to our species and the human brain in particular. Professor Lahn's team[76] examined the DNA of 214 genes involved in brain development in humans. Their work suggested that humans evolved their cognitive abilities from an enormous number of mutations occurring in a short period through an intensive selection favouring complex cognitive abilities. Our brains are disproportionately big; much bigger than the brains of our closest relatives the chimpanzees, and the conventional explanation is that the emergence of society spurred growth in intelligence.

The wider biological conclusion is that the relative size of the human brain does not fit the usual slow evolutionary trend for other organisms and organs; something dramatic must have spurred brain growth over a relatively short period. A source identified only as an unpublished NSA report concluded that sixty-four epigenetic augmentations of our genome preceded the arrival on the scene of modern humans. Could genetic enhancement over only a few generations be the mechanism which also fits these observed features of human DNA?

Michael Tellinger reported that the Human Genome Project found some 3 billion base pairs in the human genome and that the vast majority of the sequence of genes are the same for all humans, implying a fairly recent common origin. On this point, Hancock suggested that the emergence of *Homo sapiens sapiens* appeared to have occurred in the epoch 15,000–8,000 BC (i.e., during the last Ice Age). Our recent origin could perhaps explain why further speciation of human beings has not occurred. Despite superficial racial differences between us, we are all still mutually fertile, and this implies that we are a very recent species!

Approximately 97 per cent of the human DNA appears not to be coded directly as genes. As noted, this component was originally called 'junk DNA' but has now been shown to have a modifying function on the expression of adjacent genes.[45] Thus, while the DNA sequence of genes between humans and chimpanzees is nearly identical, there are large 'gaps' adjacent to human genes that affect the extent to which genes can be turned on or off. These gaps contain viral-like sequences called retrotransposons. They make up about half of the genomes of both humans and chimpanzees. A recently discovered piece of 'junk DNA' (called HACNS1) in humans might have contributed to our astonishing manual dexterity, as new research by Polavarapu and colleagues[15] suggests. In humans, this component of the genome seems to activate genes in the budding wrist and thumb. In chimp and monkey versions the same gene seems only capable of switching on in the developing shoulder. We may speculate that the importance of this piece of 'reprogramming' of the same gene converts proto-humans into a useful servant race having the needed manual dexterity.

The conclusion from all of this is still uncertain of course, but circumstantial evidence suggests that our genome has been tampered with, both in our earlier hominid evolution close to a million years ago[6] and much more recently by the Anunnaki in our evolution and early culture. The overall result of human evolution is that from a cluster of pre-hominids, only one hominid, *Homo sapiens sapiens*, survived a variety of climatic crises to the present. This could be a tribute to our evolutionary plasticity or to the relatively recent occurrence of a critical stage in our evolution. Could this be evidence that an extensive breeding program was employed on humans to produce an efficient genotype, with the radical changes just identified occurring quite recently in geological time? The rest of my document reflects established conclusions on this topic, but it also discusses what passed through my mind while thinking about this question.

Highlights of My Personal DNA

In contrast to the complications of investigating human prehistory 'on the ground' from ancient artefacts, writings, and mega-constructions, my personal antecedents came to light after sending a swab of my mouth cell debris for DNA analysis. This may seem irrelevant to a discussion of ancient human origins, but I believe that the ease of determining our personal history by DNA analysis compares favourably with what little we know, or can find out, about our real history over the last 300,000 years or so using classical archaeological methods. If we exclude the last 10,000 years where we have accumulated scattered bits of historical information, components of my personal DNA shown in table 2, with a minimum of interpretation, suggest that my ancestors' implied migrations extended far back before the historically documented era. These extensive hypothesized movements could have been in response to climatic change. However, extensive migrations through different habitats from subtropical to cold-temperate have also been documented. If selective breeding was a possibility implemented on us by a hyper-intelligent and technically proficient species, this type of movement through different environments would have been valuable in selecting a hardy final strain aimed at improvement of a single global species.

Some features of the resulting *Homo s. sapiens* genotype could later have been advantageous in competition with other species such as Neanderthal man, who had a larger cranial capacity than ours, and may have been more intelligent on average.[22] However, the Neanderthals seem to have been largely confined to family-sized communities in arcto-boreal environments where agriculture was impossible. In any case, components of their DNA were incorporated into ours, together with contributions from the mysterious Denisovans. Our net advantages were the invention of speech, the promotion of social instincts, the eventual production of food by agriculture, our high fecundity, and our efficiency in bringing up our infants due to the supportive role of grandparents. All these features aided in social cooperation, and promoted population survival and growth. The efficiency of earlier *H. sapiens* in reproduction, and our mobility and aggressive behaviour, must also have been key selective factors. These competitive advantages were not equalled by our skills in population self-control. According to Sitchin, the Anunnaki leader, Enlil, was reported to have been disgusted by the early reproductive obsession of his short-lived human workers and attempted, unsuccessfully, to ban Anunnaki-human intercourse or marriage.

I realize that the migratory history of my ancestors portrayed in table I can be regarded as partly creative, since the time intervals deduced from readings of DNA are not exact and not all genetic components participated in all migrations. Yet the fact remains that the evolution of the genetic strain leading to myself did not occur in a single place. There were successive migratory phases caused variously by climate, starvation, invasions, and conflicts, as implied by the sequential DNA interpretation shown in simplified form in this table. Apparently, my ancestors (and many of yours) passed through Mesopotamia between 60,000 and 35,000 years ago (Fig 2), which may be relevant to this essay.

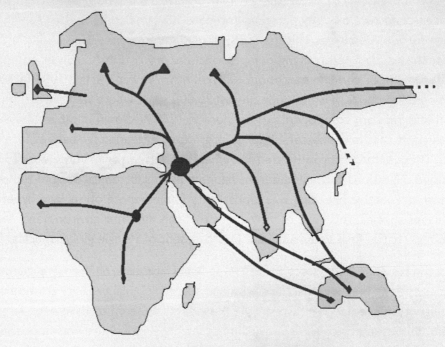

Figure 2. Y-Chromosomal co-ancestry for John F. Caddy with similar human genotypes in Eurasia-Australia imply migrations following Haplogroup R1b (dotted lines lead to locations on Pacific islands and the American landmass).

Table 1. The analysis of DNA samples from my cheek swab suggest a migratory trajectory for my ancestors from Africa across Eurasia, over a period of 300,000 years.

Y chromo-some	Patrilineal	X chromo-some	Matrilineal
	Sharing with all other men, a single patrilinear 'Y-chromo-somal Adam' >300,000 yrs ago	Haplo-groups LI-L6.	One matrilineal ancestor traced back to 150,000 years ago of hunter-gatherers, in what is now Kenya, Ethiopia, and Tanzania.
Haplo-groups A, BT	Mainly remained in NE Africa, then to sub-Saharan Africa	L3-> M,N	Left E. Africa, probably crossing the Red Sea to the SW Asian corridor from Anatolia to NW India. 60,000 years ago, this group may have been associated with Neanderthal populations in Israel. Also found in Turkey, Pakistan, and India.
Haplo-groups C and D, Then F	First groups C+D, moved out of Africa to SE Asia + Australia. Second group F was dominant 45,000 years ago, and expanded from the Middle East.		
Haplo-group K	Tribes of hunters: Middle East->SE. Asia -> north -> Himalayas-> Central Asia.	Haplo-group N	Some moved back to Africa, the Near East, and Central Asia, while my line was found in Anatolia and Mesopotamia.
MNOPS ->P	To steppes of Central Asia in the Ice Age	R	Anatolia + Near East 35,000 years ago.
RI-> MI73	Asia -> Europe	T (Clan Tara)	Moved to Europe at end of the Ice Age, with agricultural skills.

Y chromo-some	Patrilineal	X chromo-some	Matrilineal
M343 (R1b)	Cro-Magnon (Solutrean culture) cave art: coincided with the end of the Neanderthals. We moved to S. Europe in the cold period.		Esp. in Scandinavia, the Baltic and Caucasus areas, including the Udmurts, a Finnish tribe in the Urals
R1b	Dominant ancestry for most European men, especially in Ireland and southern England	T1a	I was born in Cumbria, in the NW of England, in 1940, of English, Scots, and Welsh ancestry.

The role of early conflicts between human migratory bands may have been exaggerated, but there is evidence that we exchanged genetic material with other groups of hominids we fraternized with en route. Aggression between groups of humans is common in our history, including capture, rape, and abduction of the females of our foes. This must have long been a method of genetic enrichment, particularly in light of more recent conflicts involving 'ethnic cleansing', where such an antique method of cross-breeding was a consequence of aggressive behaviour. (It is still characteristic of our species when influenced by xenophobia, nationalism, and/or religious intolerance). However, a more specific mechanism of genetic diversification also emerges from the accounts of Sitchin and is discussed in the following sections.

Evolution of Intelligent Species on Earth before Humans?

The idea of Knight and Butler that super-intelligent beings from elsewhere might find it desirable to increase the numbers of other intelligent species in the cosmos contrasts with the usual idea that extraterrestrials come to conquer. Since a technically superior species could quite easily conquer or eliminate a population living on the surface of a planet, in the light of historical facts this hypothesis is illogical. The fact that 'they' have not done this, by contrast, suggests a more benevolent or at least tolerant attitude to a species of developing intelligence; or could it be that they are infiltrating our populations by slowly modifying their genotypes with ours (or vice versa)? One reason might be to gain immunity to our microflora, implying an interest in residing on Earth, or slowly modifying ours to bring us closer to some galactic standard, implying a future role for us as disease-free members of an off-planet community?

A less comforting hypothesis emerges from recent efforts to identify planets beyond the solar system with the capability of supporting life. Of the hundreds or even thousands now identified, only one or two lie in that restricted 'Goldilocks Zone' at a distance from their star where its radiation leaves water on the planetary surface as a liquid rather than as ice or steam. This of course means that the Earth is valuable to ET for reasons that go far beyond our presence on its surface – a mature ecosystem and breathable air are rare resources in the galaxy, and approaches to planetary conquest that damage these assets would be seriously counterproductive.

The frequent kidnappings of our species reported recently[16,18] (but which may have been going on for millennia) suggest that we may have value as an off-planet labour force, but we may be still more valuable as a source of genetic material. Could genes from a young addition to the galactic sum of intelligent species have some utility for those hyper-intelligent species approaching senescence due to a loss of genetic diversity after millennia of clonations? If so, we may be valuable as a source of genes for 'reviving' small populations of ancient races with a low genetic diversity.

Visiting species of advanced intelligence may also have more altruistic motives that we do not yet understand. For example, peaceful behaviour in the galaxy could be a pre-requirement for a species wishing to avoid extinction in the long run by bombardment from space. Those who have had contact with extraterrestrials have expressed optimistic opinions, notably that a confederation of intelligent species exists in the local cosmos aimed at protecting us and other developing intelligences from aggressive exploitation by superior intelligences and technologies, until we emerge from our own phase of truculent barbarism. This optimistic view may explain, for example, certain incidents involving the deactivation of nuclear warheads by UFOs[17] and why reports on the deactivation of nuclear warheads by UFOs are common on the web.[54,69]

Given the chaos produced by the long-term geo-climatic changes on our planet discussed earlier, a rapid development of space flight may be necessary for the corporeal and informational continuity of a newly intelligent species. Is it also possible, as suggested earlier, that repetitive evolution on this planet allowed several intelligent life forms to survive in sequence in response to common ecological stimuli? Could one of these intelligent species have evolved, and already left 'our' planet, during the enormous time intervals before intelligent mammals evolved? Some limited evidence for this has already been discussed in the case of reptilian intelligence. Certainly, very rapid evolutionary changes have been documented for our own genus, *Homo,* which date back to less than a million years ago. However, as suggested here, we may have had a great deal of help in the recent evolutionary process.

It is certain that extraterrestrial species other than those mentioned here visited our planet, attracted by the unusual richness of terrestrial environments and biota and by the powerful fields of prana-like energy surrounding the earth. ETs such as the Greys and seventeen or more extraterrestrial species that have apparently visited us[28] are potentially relevant to this theme, but this account focuses mainly on the Anunnaki, where we seem to have a little more available evidence. Nonetheless, it is quite possible that modifications of terrestrial ecosystems or species could have been arranged by other entities involved in 'Cosmic Ecosystem Studies'. These could have involved seeding DNA (or bringing biota from elsewhere) or transporting useful species (including ourselves?) off planet. Certainly, a wide range of possibilities are left open by the inevitable rarity of ancient artefacts, but it would be very curious if a planet with exceptional biodiversity, a viable atmosphere, abundant water resources, and a complex biosphere, had not been visited by extraterrestrials during its billions of years of history!

The conventional reason for the sceptical scientist to exclude this possibility, is through a comforting reference to Albert Einstein's discovery of the physical limit imposed by the speed of light. This would mean, on the face of it, that a visit here from inhabited planets elsewhere in the galaxy could take thousands of years. This hasty conclusion reveals of course that the speaker a) does not wish to believe in further scientific evolution, or extensive evidence for intelligent beings coming from elsewhere. It could also be that the science skeptic is not familiar with a number of methods of greatly shortening the voyage, either through the use of cosmic 'worm holes', or simply from the application of thousands more years of research than we have been involved in.

Tellinger[11] suggested that the Anunnaki transported vegetables and four-legged farm animals (sheep were mentioned) from their planet to Earth. This seems unlikely on the face of it, but could it be checked by DNA analysis of these or other organisms for 'anomalies'? One common intelligent species with excellent colour change capabilities and a complex nervous system has been found to have a genome and many genes much different from supposedly related molluscs.[83] In fact, the most intelligent and often the largest marine invertebrate, the octopus was referred to, perhaps jokingly, by its geneticists as the 'alien'; and we may perhaps expect a growing interest in animal and plant genomes to uncover other visitors to our planet.

Other Ideas on Our 'Galactic Grooming'

In her book *Hybrid Humans: Scientific Evidence of Our 800,000-Year-Old Alien Legacy*,[6] Daniella Fenton discusses what she considered to be even earlier changes imposed by an external agency on the genes associated with human brain size, our neural structures, and information processing. She asserts that 780,000 years ago, such additions were formed from 'junk DNA', with fragments of genes cut out, copied, and reinserted. Fenton pointed to the mysterious fusion of chromosome-2 as further evidence of early extraterrestrial experimentation. This feature is found in all large-brained humans, including Neanderthals and Denisovans, but not in any other primate. She explained that from 780,000 years ago, all humans demonstrate the fusion of chromosome-2, which implies that this mutation resulted in benefits. It appeared suddenly in a considerable number of individuals and became a dominant trait. This short-term wholesale change does not fit in with the usual time course of natural mutations. Fenton used her skills as a psychic medium, aided by accounts of shamanic dreams, altered states of consciousness and past life experiences, to search for evidence of early human encounters with extraterrestrials. Converging accounts she gathered identified an early visit around 780,000 years ago by a large crystalline spaceship from the Pleiades: a star cluster considered sacred by many early human societies.

This spaceship was shot out of orbit using advanced weaponry by a reptilian race that, according to her account, had evolved here. Destruction of the ship by the reptilians, who were then presumed to be dominant on Earth, led to a major loss of life and was seen by other highly evolved extraterrestrials in the galaxy as an unwarranted measure. She reported that the few Pleiadean survivors dedicated themselves to performing early genetic modifications on pre-humans even before our divergence into what became human, Neanderthal, and Denisovan subspecies. She speculated on the details of the genetic modifications to pre-humans they carried out. She looked for ancillary evidence for the disaster and referred to the debris of glass fragments found throughout Australia and SE Asia (called by geologists 'strewn fields of tektite'), which are the extensive scattered remains on the earth's surface she hypothesized resulted, in this case, from a high-speed impact with a glass spaceship, leaving no crater. According to Fenton's sources, after notification of this tragedy a decree was issued by a confederation of extraterrestrial civilizations which sent a battle group of what she called Leonine humanoids into Earth's orbit. The remaining reptilians, who were located in underground caverns, were ordered to vacate the earth immediately. Aerial bombardment of their strongholds by space rocks followed signs of reluctance by the reptilians to obey this decree. Taking this sequence of events seriously may offer one of the few hypotheses that we have that could explain how human beings later evolved outside the domination by earlier evolved reptilians. In the more 'imaginative' literature on 'aliens', the reptilians have become the typecast species hostile to humans—even though this seems to be in conflict with early religious beliefs in wise reptilian gods.

Fenton made efforts to identify factual information that could verify her earlier conclusions, and independent research on the human genotype she cited shows that the human genome (as for other terrestrial species) has a structure with intrinsically logical features characteristic of artificial design, and not just a result of random evolutionary processes. Shcherbak and Makukov[18] asserted this conclusion as a result of their mathematical investigations into the structure of DNA spirals. They assert that the underlying mathematical features of genotypes mean that their form could not be completely a result of natural selection but had a logical format basically reflecting an artificial origin. In other words, this hypothesis supposes that DNA was designed by an advanced intelligence billions of years before life on Earth began. Hence, this also implied panspermia.

An even more radical implication of the work of these authors is that the underlying structure of DNA is equivalent to that of a symbolic language. The message it contains is for the moment unavailable, but quite possibly, its writing preceded the Pleiadean or Anunnaki civilizations. If this conclusion is valid, we should ask; what could have been the equivalent informational and reproductive structure of the 'Inventors of DNA'? Evidently, it would be logically inconvenient if these Inventors also used DNA for transmission of their own personal structural information—and if this, in turn, was artificial! If so, this question would have to be addressed again to the Inventors' Originators! The question of Who or What was originally involved, and from Where (if an active role of the deity itself is to be excluded), is anybody's guess! The arguments that Sitchin and Fenton make on the basis of interventions by specific ETs could, of course, apply to changes made by other ETs to our genome without invalidating the basic argument that such changes *were* made!

Human Linguistic/Cultural Divergence and 'The Tower of Babel'

The gene FOX2 is present in both chimpanzees and humans and has been implicated in the human development of speech, although the human gene is slightly different chemically from that in the chimp where verbalizations are limited. Nonetheless, its importance for communication and intelligent behaviour, without necessarily involving a language, was demonstrated by the superior performance of mice running a maze after they had been transplanted with this human gene. Improved self-location persisted as long as visual clues remained in place from previous tests that they could visualize and learn from in successive trials.[27]

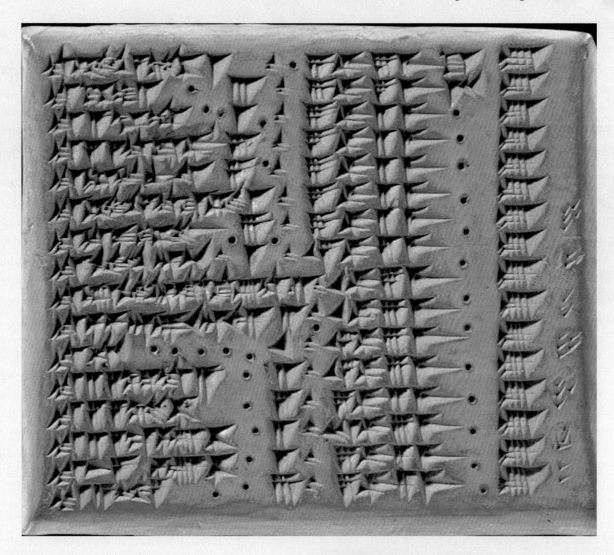

Figure 3. Sumerian cuneiform writing: an example from the Facebook page of ancient civilizations, giving the constellation names and distances between stars. (Tablet dating from the 3rd-2nd Century BC - Uruk, Iraq, in the Pergamonmuseum - Berlin, Germany).

The Sumerian language (figure 3) has been found to be largely unrelated to other tongues such as the Semitic family of languages, and given that their first written inscriptions preceded extensive scripts in other tongues, it seems logical to consider a possible unusual influence. After the upgrading by the Anunnaki of earlier proto-humans to human intelligence mentioned later in the book, it seems reasonable to ask whether the origin or structure of the Sumerian language reflects in simplified form, that spoken by the Anunnaki. This would suggest that the ancient Sumerian language had extraterrestrial roots, since it would have been used by their new masters for training their servants.

The biblical fable of the Tower of Babel seems to closely resemble a Sumerian precedent, as do other biblical accounts in 'Genesis'. In this specific case, the Anunnaki expressed concern when humans began independently to construct a ziggurat. This appears to have been related to a practical utility of these structures. Apart from the ever-present religious implications of all the mysterious structures appearing at that time for early humans, the main utility of ziggurats was supposedly their use as beacons marking landing sites for spaceships. On the subject of space travel however, the Anunnaki were determined not to allow the transfer of their advanced technologies to humans. The solution to this problem, as mentioned, resulted in the diversification of human languages.

This occurred as a consequence of their transporting groups of modified humans to distant parts of the world, to isolate them culturally. For a short-lived species, i.e. *H. sapiens*, in no more than a dozen generations (roughly 300 years), their languages would inevitably have diverged, and their customs and mutual comprehension would have deteriorated. Over the long term, a very long lived species would have used this as a practical measure for reducing interregional communication of our species, and it has continually promoted wars and misunderstandings between us. The Anunnaki were undoubtedly familiar with this 'divide and rule' principle! Different cultures were thus formed, but these retarded our technological progress and reduced competitiveness with our Anunnaki masters.

Historical Confusion between Spiritual Beings and Physical Entities from the Sky

As noted by Evans,[13] Sitchin reinterpreted certain conventional meanings recognized by nineteenth- and early twentieth-century translators of Sumerian clay tablets. Earlier on, translators had assumed the Anunnaki to be mythological or spiritual beings; however, the events described on the tablets suggested to Sitchin a contemporary realism and technological meaning that earlier translators some half century earlier would have been unaware of. (i.e., he changed some verses from a spiritual or metaphorical significance to a simpler actuality). Was this a forbidden procedure? It certainly explains why there are significant differences between his interpretation of events and that of the 'classical Sumerian school', which usually considered these accounts spiritual or mythological (i.e., Sitchin's interpretation was that the Sumerian scribes also recorded factual, and not just mythical, events about the Anunnaki).

The use of the 'god terminology' is a well-established convention among technically unsophisticated peoples, especially those with polytheistic beliefs. They often use such terms to describe superior beings arriving from somewhere beyond their geographic range of knowledge. In this instance, Sitchin interpreted the clay tablets as showing how the Sumerians greatly respected, and even worshipped, the physically larger, longer lived, and more civilized hominids who arrived on Earth to guide them, and how the Anunnaki encouraged this attitude. A reason for awe and respect arose from the way they apparently led humans into a stratified urban society supported by a previously unfamiliar system of irrigated agriculture. From Sumerian accounts, it seems that the term 'god' was used by the Anunnaki as the equivalent of 'chief' for their kings, and the Sumerians used it as a general term for the visiting species.

During the Second World War, American planes regularly flew over the hilltop terrain of New Guinea tribes, which had yet to be contacted by modern civilization. This provoked a curious phenomenon: a model of an aeroplane was constructed of bamboo by the natives and placed on a hilltop. It was apparently intended to attract its potential 'mate' from the sky above, who would also bring gifts for the tribe (The 'Cargo Cult'[28]). I mention this because the problem early humanity encountered in distinguishing sky visitors from gods persists today. Not surprisingly then, the Anunnaki, who arrived from the sky, were referred to as 'gods' by the Sumerians and also by their successors, the Babylonians, who used Sumerian inscriptions for spiritual guidance much as we have used the Bible. The theological nomenclature used on the tablets convinced earlier translators that the visitors were spiritual entities. Not surprising, perhaps, is the early reluctance of these translators to suppose that extraterrestrials exist. For example, in *The Ark before Noah*, Irving Finkel[29] wrote of the Babylonian kings that 'they professed themselves constantly under the protection from the most powerful gods', such as Enlil, who we will meet later through Sitchin as an apparently real entity; he was reported to have been the leader of the Anunnaki during the early age of the Sumerians.

Such an automatic 'conversion to divinity' by the early translators of strange entities on Sumerian and post-Sumerian era tablets is understandable, since, in the late nineteenth century, the translators assumed that to the Sumerians, religion was culturally dominant in a technically primitive ancient world. It was also perhaps because the nineteenth-century translators lived before the invention of flying machines! Hence, a statement such as 'Enlil looked down from the sky' obviously implied to these early translators the spiritual presence of a deity looking down from the heavens, and not a physical body in a flying machine.

The fact that the Anunnaki were reported to be long-lived, twice as tall as human beings at twelve feet or more in height, with elongated crania and red hair, must have made them seem godlike to early humans. Thus, as noted in chapter 6 of the Book of Genesis, 'There were giants in the earth in those days'. It must also be a feature of repeated oral transmission of information over long periods that key figures in history begin to assume divine proportions. In subsequent regional civilizations based on Sumerian precedents, it would not be surprising then that the Akkadians and Babylonians, who borrowed extensively from the Sumerian literature, accorded the Anunnaki truly divine characteristics.

Religious Beliefs: Mythologies or Ancient Events?

Hancock's idea was that we should look for messages in ancient artefacts, traditions, and mythologies, and that the 'creation myths' in many cultures may contain kernels of factual information on our origins. Even a superficial examination of early religious paraphernalia, for example, shows that early mankind reserved a prominent place for serpents in their religious beliefs. They occur on royal headdresses in India, Egypt, and Mexico. Giant wise serpents are encountered in ayahuasca trances. Enormous feathered serpents were believed to be responsible for the creation of the earth in the Australian 'dream time' of aboriginal traditions. The plumed serpent in Mexico, the dragons in China, and rattlesnake portrayals on Mayan pyramids are other examples that indicate our species' early instinctive reverence for a once all-powerful reptilian race. This, of course, is also evident in the story of the Garden of Eden, where the snake in the apple tree was all-knowing. All this may be a consequence, then, of a far distant age, when a reptilian race was dominant on this planet and was not necessarily regarded with fear and hostility by early humans.

According to Maltese archaeologists, it may be relevant that the religion of the serpent was followed by an ancient race that first occupied the fertile half-moon area (particularly Anatolia and Kurdistan) and Egypt [94], following migrations dating back to 6000–4000 BC. This race (of cone-heads, as they were referred to elsewhere in this text) even reached Malta, where they apparently disappeared around 2500 BC. One can also imagine that 'snakes' would be a common term by humans for a hypothetical race of intelligent reptilians that earlier in our evolution we may have had dealings with. If we look further back in time, the door was left open over several hundred million years for intelligent reptilians to complete their evolution on Earth, and evidence for one of their survivors within historical times has recently been reported.[9,14]

Mention could be made at this point of the elongated headdresses of the pharaohs (e.g. figure 3), and the actual head profile of the Egyptian pharaoh Akhenaton[30] and his children, whose bizarre body shapes and elongated 'cone skulls' have been commented on extensively. An example of such a skull configuration is shown in figure 3. (The tall mitre of the pope in Rome and feather headdresses of American Indian tribes both suggest that it was not only early Christians who showed this high respect for apparently elongated heads!)

Apart from his resemblance to the giant humanoids associated with the pre-Pharaonic stage of Egyptian, Sumerian, and pre-Incan civilizations and their megalithic constructions elsewhere,[31] the new religious concepts Akhenaton introduced are indirect evidence that he was influenced by a radically different belief system than found in Egypt (and possibly elsewhere on Earth) at that time. This is the first documented occurrence of monotheism we are aware of in world history, and it is usually considered to be his individual act of personal inspiration in ascribing divine intelligence to a supreme cosmic entity. This supreme entity was certainly located off planet and was assumed to be the sun. (An alternative proposal is that it could have been the king of Nibiru, who was considered by the Anunnaki to have godlike characteristics in their apparently rigid theocracy). However, although ceremonies for the unknown cult he proposed were carried out in Egypt, he may have acquired this belief system and its associated ceremonies from somewhere else and tried (in vain) to introduce them to a land where polytheism had long been the norm. Polytheism continued to be so after his reign, when monumental evidence of his monotheistic beliefs were erased by the revival of the old religion. We may note that this appears to be the first example of a 'sky-based' religion introduced to humans by Akhenaton, and a close relationship between Akhenaton and the prophet Moses illustrates the possible linkage of some of his beliefs with those of the early Israelites.

The archaeologist Walter B. Emery,[31] recounts in his book *Archaic Egypt* how he found skeletal remains of large individuals in tombs from pre-dynastic times in Upper Egypt. These had skulls of an abnormal size: 25 per cent longer than wide, with the normal skull sutures absent in many. He did not hesitate to identify them with the 'Followers of Horus'.

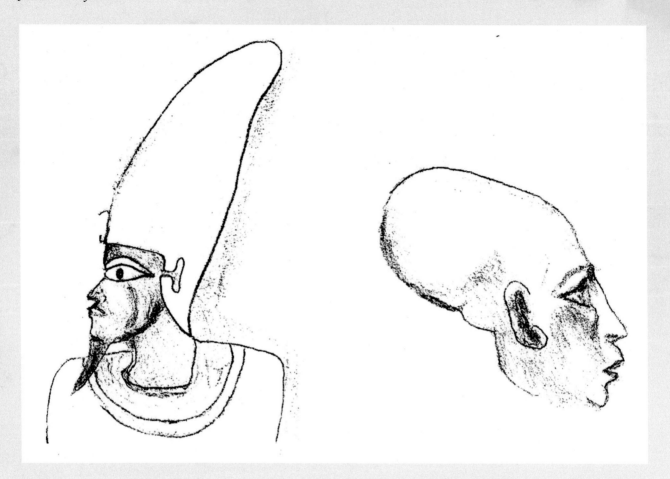

Figure 4. An early pharaoh, illustrating the exaggerated height of his royal crown, and (right) a portrayal of an ancient Egyptian with an elongated skull (drawn from illustrations on 'Ancient Pages').

Emery supposed that these giant individuals had originally filled an important priestly role. With regard to this race with long-headed skulls, he noted that this lineage seemed to come from a cycle of civilization dating from before the Flood, which we can tentatively identify as that of the Anunnaki or their hybrids with humans. Much later, the Pharaoh Akhenaton was portrayed in his statues and bas-reliefs (and also with his royal family) as an individual with a long head and a serpent likeness that Emery mentioned was characteristic of the pre-dynastic Egyptian rulers. According to some experts, these represented the biblical giants called the Nephilim, and were close in appearance to the elongated-skulled individuals found in an underground temple in Malta.[77] The crania of the Malta skulls were reported to be practically identical with the X-rays of Tuthankamon, Akhenaton's son, showing a dolichocephalous cranium. Substantially, it seems that the Maltese craniums are the relics (although archaeologically still misunderstood) of a sacerdotal race that survived in Egypt and Malta from archaic ages until at least 2,500 B.C.

It was during the period of the 'Followers of Horus' (from 4,000 BC) that the Pharaonic dynasties began and gave rise to the unequalled degree of sophistication of the early Egyptian civilization. One quotation describing the work of the archaeologist, professor Timmerman, is as follows: 'In 1881, when professor Timmerman was exploring the ruins of an ancient temple of Isis … 16 miles below Najar Djfard, he opened a row of tombs in which some prehistoric race of giants had been buried. One measured seven feet and eight inches in length and the largest eleven feet one inch. It was believed that the tombs dated back to the year 1043 BC.' This observation seems to suggest that the Anunnaki, or Anunnaki hybrids, played an important role in establishing the Pharaonic dynasties, and this could also explain the longevity of the early pharaohs.

Similar elongated skulls have been found in several world regions[102].

For example, Von Tschudi and Rivero identified three pre-Inca dolichocephalic races in Peru. Of these, the Chinchas had a lengthened skull caused by bandaging their children's skulls so as to resemble members of the other two tribes, the Aymaras and Huancas, who had naturally elongated skulls. These last two groups[50] preceded the Incas and were reportedly subsequently influential on both the Incas and the Mayans. It is interesting, by the way, to note that the oldest pre-Inca city, Tiahuanaco, dates from the same period as pre-dynastic Egypt and shows similar features, such as their megalithic construction techniques, perhaps implying some communication between them. It also seems reasonable to assert that these South American residents were members of an antediluvian race with a naturally elongated conical skull, examples of which have been found all over the world. Skulls from this period can be seen in the Tihuanaco Museum and are illustrated on the Internet. It is therefore asserted that there once existed an antediluvian race with a naturally elongated skull. This is affirmed by researchers such as Dr Tschudi, who possesses a fossil from that time of a seven-month-old foetus with a dolichocephalic skull found in the womb of its mother; this shows that natural reproduction of what we have called the Nephilim, or cone heads, was possible.

One characteristic of exhibits of these 'cone-head giants' and other deviants from the current human skeletal norm, is how these dolichocephalic skulls have been withdrawn from public view, as in the Valetta museum of Malta, to avoid offending the sensitivities of some visitors. A parallel process, for the same reason perhaps, was mentioned for a famous American museum? Thus, of the seven hundred examples found on first opening the hypogeum of Hal Saflieni and associated tombs of the megalithic temples on the island, few remain for inspection. Of the skulls found in the Hal Saflienti Hypogeum, a lengthened posterior part of the skullcap was evident.[38] No analogous pathological cases were found in the medical literature by an investigation carried out at the time. The international medical literature showed no natural cases of a lack of median knitting of the 'sagitta' shown by these skulls. This anomaly emphasized the difference between a natural elongation of the cranium and that due to bandaging an infant's skull between boards as in pre-Colombian civilizations (Fig 5). On this basis, the archaeologists involved proposed that the skulls found in the Hypogeum represented a group of peoples who had a natural genetic tendency for elongated skulls and were integrally involved in the activities of the temple builders of that time. This race appears to have been devoted to the priesthood, the worship of a serpent god, and the building of megalithic monuments.

Gigal[50] indicated there was probably a migration of long-headed people from Egypt to Malta, and traces of them are also found in the Cretan civilization. This race seems to have been devoted to the priesthood and teaching, and to one other point in common everywhere: the building of megalithic monuments. How can we explain this other than to say that these people were the descendants of an antediluvian race, the Anunnaki or their hybrids, discussed elsewhere in this report?

Fig 5. Tall heads symbolize high status: a) Woman of the Chinook tribe with her baby being subject to skull elongation b) Ancient Japanese god c) Busby of highland foot soldier d) Upper class representative with a top hat e) The mitre on a bishop.

An observation made on the people of pre-dynastic Egypt was that they were beginning to lose their subtle senses[31] and the 'divine beings' present at the time (presumably the tall, cone-headed priests) attempted to halt this loss of extrasensory perceptions. One wonders whether highly developed extrasensory perceptions (ESP) or telekinesis in this priestly caste was linked in some way to their unusual architectural skills? Their 'magical' skills may also provide a clue to their subsequent disappearance. Due to their enhanced powers of intellect and 'magic powers', were they later subject to aggression from their less gifted human subjects in response to previous mistreatments? Did such a hypothetical interracial conflict lead to the advanced symptoms of warfare which left vitrified traces on the ground, as mentioned later in this report?

Sumerian Sources, Mythology, and Genesis

Nonetheless, the complex political and ethnic events following the decline of the Roman Empire are now being deduced in part from genetic analysis[97]. One example comes from the remains of thirteen women discovered in medieval Bavarian burials that originated from what Is now Bulgaria and Romania. Their elongated skulls and darker eyes and hair were genetic features contrasting with a local population of blue-eyed blondes.They shared a genetic ancestry with southeastern Europeans, with a minor contribution from Central Asia where the tribes of the Huns were mentioned. It seems possible that they held elite status since they were often buried with their jewellery, and may have been traded as brides for diplomatic reasons. One showed Central Asian ancestry, and could have migrated from the Black Sea region to Bavaria around the year 500. Perhaps these long-headed brides were produced by cranial deformation in imitation of the Anunnaki, in proximity to one of the areas of dominance of their cone-headed hybrids?

Graham Hancock, and Christopher Knight working with Graham Butler, were among the authors who independently postulated the existence of an advanced pre-Ice Age civilization with a level of sophistication in some respects superior to ours. They supposed that the Sumerian and Egyptian civilizations did not develop 'from scratch' our knowledge of astronomy or the system of numbering and measures we have inherited. In acquiring these and other skills, we may have been beneficiaries of a 'First Civilization'. This leaves open the question of whether this First Civilization was of another more advanced species?

The ideas expressed in Knight and Butler's books stem from practical measurements of the dimensions of Stonehenge and other Neolithic ceremonial sites surveyed by Alexander Thom,[32] who showed that their construction was based on a common unit of measurement christened the 'megalithic yard' (83 cm). Knight and Butler demonstrated by simple measurements involving containers of this ancient measure, that this length is the basis for our common measures of weight (pounds), volume (pints), and—using a pendulum of a given measure—time (minutes and seconds). This in itself is an astounding illustration of the antiquity of the common measures we use daily, and the reader is referred to their books for more details. They also showed how these measures relate to the geostatistical dimensions of the earth and solar system when they were expressed in megalithic yards. Remarkably, when converted to megalithic yards these cosmic measurements often yielded large whole numbers (e.g., 60 million), supposedly showing that the megalithic yard was derived from cosmic measurements and not vice versa, and that our 'common or garden measures' were based on a prior knowledge of the dimensions of the solar system, and other measures equally improbable at first sight, such as the speed of light! The implication of these discoveries of course is that the originators of these measures were a species with considerable experience of space travel [84].

Nibiru: The Anunnaki Planet

Although the idea of an unknown planet in our solar system is difficult to credit, Nibiru was described by the Sumerians as a radiant planet with a reddish hue and referred to it as 'the planet of crossing' since when returning within the inner solar system, its orbit crossed between Mars and Jupiter. There have been several hypotheses by astronomers seeking to explain 'anomalies' in the orbits of distant solar system planets as due to the gravitational influence of a large undiscovered 'Planet 9' further out in space [89]. This question is relevant and apparently still unresolved. According to Zecharia Sitchin, the Anunnaki (see e.g., figure 4), came to Earth almost half a million years ago from the planet Nibiru when it entered the solar system. Avoiding premature scepticism on an unresolved issue, the strange elliptical orbit of Nibiru suggests that it is planet somewhat larger than the earth closely rotating around a dwarf star, both captured gravitationally by the sun. This resulted in an orbital overshoot taking them far outside the regular orbits of other planets in the solar system. Examining the Anunnaki symbol for Nibiru does not exclude this hypothesis (see figure 13), given that this symbol could be interpreted to show a small radiant body in front of a relatively large planet somewhat larger than the Earth. In fact, a recent article in the *New Scientist*[85] pointed out that potentially habitable worlds are more common around cooler small red dwarfs and that they rotate much closer to the star than the earth does to the sun. Hence, some astronomers have switched their attention to red dwarf stars in their search for habitable planets.

Sitchin's interpretation of the Sumerian clay tablets in his book *The Twelfth Planet*, referred to this astral body as planet X, or the 'ninth planet'. These last two terms were used by astronomers such as Alessandro Morbidelli,[36] who suggested that there may be a larger planet beyond Pluto. Modern astronomers claim to have found evidence for the existence of such a 'hidden' planet whose orbit is currently outside Pluto.

This deduction is based on its influence on the orbits of other planets of our solar system. Sumerian tablets describe the planet, somewhat larger than the earth, radiating orange-red light but protected by a thick atmosphere. Ancient Egyptians and Babylonians also described this planet (see Christos Djonis[37]), referring to it as the 'Planet of the Crossing', since its orbit within the solar system passed through those of other planets, on occasions with disastrous results. As described in the analyses of Sitchin, this planet's unusually elliptical orbit takes it past Neptune and Pluto, and it long remains far from the centre of the solar system, with an orbital duration believed to repeat every 3,600 years. The orbit of Nibiru may be envisaged as similar to that of the comets, which spend a great proportion of their long orbits outside the planets of the inner solar system. Quite a number of questions about this 'planet' remain unanswered: Is it associated with a dwarf star? What are its radiation wavelengths that would allow photosynthesis? Presumably wavelengths similar to those arriving on Earth, otherwise the Anunnaki might have had radical problems in adapting to life here?

According to Tellinger, gravitational disturbances from a close passage of Nibiru 155,000 years ago peeled away much of the surface strata of the planet Mars. An additional effect may have been the creation of proto-comets with similar elliptical orbits to Nibiru; only over long periods do these re-enter the central solar system. Nibiru's passage through the centre of the solar system may have even accentuated the effect of the Great Flood and led either to the slippage southwards of the Antarctic continent or to the movement of our planet's poles of rotation. Either way, the phenomenon of this cosmic entity passing through the inner solar system may have resulted in a continent with a cold-temperate climate being displaced latitudinally down to the South Pole.

Ancient Writings as Evidence

Inevitably, many experts disagree with some of Sitchin's interpretations of the records inscribed on Sumerian clay tablets, but it is ironic that these Sumerian clay tablets provide us with more extensive records of events in prehistory than later civilizations which used more fragile parchment or paper records. Nevertheless, these records are the source of controversy, along with aspects of the interpretation placed on them by Professor Sitchin. Many of the later more fragile records on papyrus or parchment were lost in the destruction of the Egyptian library in Alexandria. Nonetheless, the Sumerian records and the explosive conclusions drawn by Sitchin are often ignored. Perhaps this is because it is usually assumed that they represent mythical events; however, from more recent discoveries I believe that his literal translations provide the basis for a coherent explanation.

'Anunnaki' was originally translated from Sumerian tablets as 'superior beings' or 'gods', but a more practical interpretation which takes Sitchin's work into account is that they were advanced extraterrestrials from off-planet. It is difficult to identify the origin of unusual hominids, and DNA analysis of what may be the skeletal remains of Anunnaki or their human hybrids is still preliminary.

Sumerian texts describe Enki (in his later appellation as the god Ea) as one of two half-brothers, both sons of Anu the king of the planet Nibiru. Enki was apparently the person who gave the first humans 'the fruits of the tree of knowledge'. The Epic of Gilgamesh also describes how Ea was merciful and delivered instructions to Utnapishtim (Noah in the biblical version) on how to build a ship and rescue a small group of humans and their cultivated 'seeds' from the flood: 'The gods were angry at mankind so they sent a flood to destroy him. The god Ea warned Utnapishtim and instructed him to build an enormous boat to save himself, his family, and "the seed of all living things".

Enki was reported to be more humane in his treatment of his modified human servants than Enlil, who was his half-brother and later replaced him as the ruler on Earth. As our distant benefactor, Enki supposedly passed on his teachings to this new race to explain their future role under the Anunnaki. It is strange though that Enki became identified with the snake in the Garden of Eden, given that with their imposing hairstyles and beards, the Anunnaki were closer to mammals than to reptiles. One view is that sinister influences were at work behind this description, and Tellinger suggested that identifying Enki in this way was part of a smear campaign carried out by Enlil to discredit Enki. An alternative explanation could be that the cone-heads revered the snake god as the most intelligent of deities and not as a representative of evil. This interpretation seems to fit well with the snake being revered by other early peoples, as in the case of the Rainbow Serpent of the aborigines of Australia and the Plumed Serpent in ancient Mexico?

The Arrival of the Anunnaki

From the clay tablets translated by Sitchin, fifty of the Annunaki under the leadership of Enki were reported to have made the original landing in the Indian Ocean and waded ashore in a marsh, probably on the estuaries of the Tigris or Euphrates Rivers.[1] As Sitchin reported from Sumerian clay tablets, these marshes were drained shortly afterwards for agricultural purposes. Proto-humans were later used as a labour force after modifying them to increase their intelligence, to facilitate their use of language, their ability to receive orders, and their effectiveness as workers. They were also used as scribes. In their much earlier version of the Garden of Eden mythology, the Sumerians implied that their own race originated as proto-humans and named themselves the 'Created Ones' (i.e., they described in simple terms that the technologically superior Anunnaki modified them through what we would call gene insertion and/or cross-breeding). After a very long period of Anunnaki rule, and subsequent rule by hybrid Anunnaki-humans, they were eventually considered fit to take over as efficient rulers of their later Mesopotamian cities.

The Garden of Eden and the 'Upgrading' of Proto-Humans for Work Duty

According to Sitchin, the Sumerian tablets described the first chief of the Anunnaki on Earth as Enki, a member of the ruling Nephilim caste who was also a competent scientist. It was reported that it was Enki who first proposed to the king the idea of modifying and upgrading a native hominid species to replace the rebelling Anunnaki workmen in the African goldmines. The Anunnaki workmen had objected to the heat and discomfort in the mines, had laid down tools, and were seriously compromising the achievement of the main mission of the Anunnaki on Earth—namely, to mine gold. To solve this social problem, King Anu arrived from Nibiru and replaced Enki as the person in charge of Earth operations by his half-brother, Enlil. This was meant as a criticism of Enki who was seen to be responsible for the slowdown in gold production.

This downgrading of Enki led to much strife on Earth later on. Nonetheless, Enki was permitted to proceed with his experiments in human genetics helped by Ninhursag, daughter of King Anu. Ninhursag was already responsible for the Shuruppak, the Nephilim medical center, and here, according to Sitchin, they modified the proto-humans by inserting splices of the Anunnaki genome into their DNA to make them useful for manual work and later as scribes to record the abundant information they left us. Sumerian accounts suggest that the modifications made to the primitive humans helped them to think logically but more importantly, to follow orders. Thus, a medical/genetic intervention made modified hominids valuable as a workforce for their more intellectually and technologically advanced masters.

The biblical story of the Garden of Eden closely resembles an earlier Sumerian story of a similar encounter arranged by a high authority. As mentioned, Enki was originally the chief of the Anunnaki on Earth and also the person responsible for the genetic upgrading of the proto-humans. As the scientist responsible, his role in the Sumerian story was effectively that of the all-knowing serpent described in the story in Genesis. In Sitchin's interpretation of the Sumerian record our acquisition of the gift of reason may be an allegory, recording an event in prehistory that simply involved our education by a more intelligent species. An alternative interpretation follows from the political rift between the followers of Enki and those of Enlil, who took over from him as the leader of the Anunnaki on Earth. Identifying Enki as a serpent could have been a way of downgrading him in the eyes of the new humans.

The earlier Sumerian version of this fable according to Zecharia Sitchin, was a mythology developed by humanity while the Anunnaki reigned on Earth with modified human beings as their slaves. That slavery played a key role in our early history is not a popular hypothesis, but it allowed our species to learn by first hand observation from their masters. Previous authors (Sitchin[1] and Marlene Evans[13]) agree that the Anunnaki genetically engineered an early hominid. into an intelligent species we would recognize as *Homo sapiens* for work in their plantations, and later as scribes. Their main task was as labourers in the southern African goldmines. The Sumerian tablets also described how mankind was used by their 'gods' to tend food items in the royal gardens for their masters' consumption.

Were There Intelligent Precursors on 'Our' Planet?

The Scriptures refer to one race of people from early biblical days as the Nephilim. These were giants, referred to in the Book of Enoch as: 'The violent superhuman offspring produced when wicked angels mated with human women in the days of Noah: a race that came to dominate the antediluvian world'. In the biblical literature they are described as having the following characteristics:

- Their height was two or three times that of normal men.

- They were associated with some kind of unholy intermixing with humans before the Flood.

- In one case they are described as being polydactyl (with extra fingers and toes).

This brings us to a hypothesis: Were the Nephilim related in some way to the Anunnaki's hybrid offspring? I suppose so, and from now on in this text we distinguish between the upgraded humans and the giant progeny resulting from conventional hybridization between the Anunnaki and upgraded female humans. These latter hybrid offspring will be referred to as 'cone-heads', or by the biblical term Nephilim, and are different from the upgraded humans themselves. The Nephilim were inevitably much higher in the rigid Anunnaki social pyramid than upgraded humans. The normally dimensioned offspring of the pre-humans resulting from Enki's genetic experiments were our ancestors, but not the hybrid cone-headed giants.

Figure 6: Sketch of an Anunnaki king portrayed with wings, not as bodily features but symbolizing his power to travel above the earth. Note also the 'watch' on his right wrist and the 'handbag' in his left hand. As mentioned later, these mysterious possessions were seen in many of the portrayals of Anunnaki distant from Sumer (image drawn from www.pinterest. it). The firmly clutched bag has been speculated to be similar in function to remote vehicle control equipment, assuming he arrived at distant destinations by spacecraft?

Although the character traits of the modified humans produced by this operation were not specified, they may be described by a well-established duality of mental types, reflecting the difference in behavior between personalities with a dominant left or dominant right hemisphere of the brain [87]. Having a dominant left hemisphere, the personality is often typified by the following character traits quite often found in military or bureaucratic functionaries:

- Responds to verbal instructions
- Problem solves logically and sequentially
- Is planned and structured
- Prefers established information
- Controls feelings
- Prefers ranked authority structures
- Draws on previously accumulated, organized information

However, a person with a dominant right hemisphere is intuitive, and tends to have the following character traits, considered typical of an artist:

- Only responds to demonstrated instructions
- Problem solves with hunches, seeing patterns/configurations
- Looks at similarities but is fluid and spontaneous
- Prefers drawing and manipulating objects
- Prefers open ended questions
- Free with feelings
- Prefers collegial authority structures
- Draws on unbounded qualitative patterns that cluster around images

One can imagine that the first of these categories, notably individuals with a dominant left hemisphere, would be the ideal servant mentality sought by the Anannuki. Whether this was a novel configuration at the time of the genetic modification of humans is hard to establish, but we can see that while the second category best reflects the flexible intuitive mental attitude required in a small foraging band, the first category would not be of much use in a small band of foragers where individuality and intuition are a priority.

Fig 7. Images copied from Jaynes[87] of Fig 7a, an Anunnaki god from an early temple in Mesopotamia, illustrating the enlarged eyes in this portrayal, which Jaynes believed were intended to induce a trance in the congregation during a sermon. This he supposed, was intended to 'program' the congregation with the views or commands expressed by the priest as the direct wishes of the god. Fig 7; A hand-held idol, also with piercing eyes was used to help the congregation to remember the commands heard in the temple, when at home or when working in the fields.

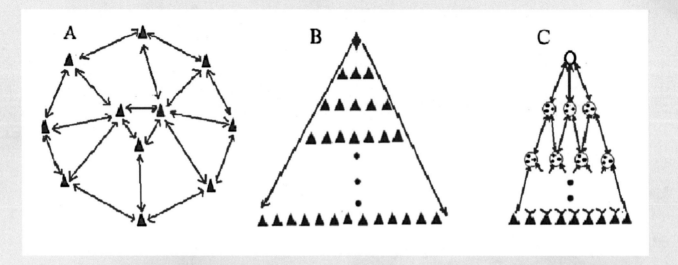

Fig 8. Three hypothesized social structures: A/ Where members of a small community hear their leader's commands, make their views known to their leader and may replace him/her. B/ The pyramidal structure of a simple totalitarian system where orders come from above and are obeyed without question. C/ A system comprising elements of A and B in a large vertically organized society, but allowing feedback on orders received by 'committees' at different levels having local or specialized knowledge.

Left brain dominance would be very useful in a large stratified community with clear lines of authority such as agricultural workers, soldiers, and those employed at the base of a chain of command. Since we may assume that the Anunnaki introduced humanity to life in organized sedentary communities as slave labour, a dominant left hemisphere may have been promoted by their genetic manipulations.

The left brain is the so-called 'masculine' hemisphere that deals with logic, analytical thought, science and math [89]. Persons so affected may also have an overactive reptilian brain: that lower section of the human brain that deals with survival instincts. In contrast, the right or 'feminine' hemisphere deals with creativity, holistic thought, spiritual awareness, compassion and intuitive skills. When the two halves of the brain are imbalanced, dangerous side effects such as psychopathic or sociopathic behaviors may result. Psychologists generally agree that psychopaths and sociopaths like to manipulate and control others, and tend to suppress their emotions while doing so. Psychopaths are generally left brain dominant and seek high positions of power, as in law, politics, religion or finance. Once they arrive there, they are often referred to as 'The Elites'. These rely on logic and science, and once they arrive high in the societal pyramid, they believe they have more rights than other people. Left brain dominant people often lack creativity, empathy, holistic understanding and spiritual awareness. In contrast, the danger of a dominant right brain occurs when people are very religious [89]; i.e., looking to a higher spiritual authority to look after them. Right brain dominance can also lead to laziness, where those so affected don't take personal responsibility for their actions and education, and hence it is easy for the Elites to use religion or politics to control them. The provisional conclusion is that the Anunnaki interventions favored left brain dominance where those affected accepted orders from above with few questions. This is as opposed to discussions round the fire and optimum foraging behavior in hunter-gatherer communities, which must have favored right brain functions such as intuition and a well-developed spatial sensibility. Phenomena such as social stratification in large, multi-level societies where decisions made at the top filter down with automatic acquiescence at each level, seem to require left brain dominance.

It came to mind today that Julian Jaynes [87] had found archaeological evidence for a mechanism developed very early on in the development of settled communities that seems to favor the evolution of what became the dominant left brain behavioral pattern. He recognized that in early Mesopotamian societies, the conscious minds of the workers were still in a rudimentary state which he referred to as 'bicameral'. Here, commands in the voice of the priest or leader were stored in one half of the brain, and transmitted to the other half where they were heard in the voice of their priest or superior, to be automatically acted upon without question. These verbal instructions were programmed during temple ceremonies where in his view a form of mass hypnotism was applied using the statues in the temples of the Anunnaki. Here the statues of the gods were portrayed with artificially big eyes (Fig 5a), which were the focus of attention during the ceremony, while the workers were being told by the priest what were their tasks for the coming days. Once at the work place, these tasks were carried out in a hallucinatory state with the repeated mental commands of the priest resounding in their heads. One supplementary aid was found in excavations of dwellings dating from around 5,600 BC. Jaynes' hypothesis was that these small statues were carried by workers and when viewed, reintroduced the earlier hypnotic state by helping the worker recall the verbal commands heard in the temple. This hypothesis of Jaynes seems to illustrate how the dominant left-brain mechanism might have been first introduced, then reinforced. Such a mechanism might be used in recent times, and contributes to our understanding of the success of later tyrants such as Hitler, Stalin, and their numerous imitators. As a substitute for a hand-held statue of their leader, humanity has discovered modern equivalents, such as hand-held telephones, television and state-controlled internet which pass on the tyrant's instructions to the workers as propaganda spelling out how they should act and vote in the sacred name of 'patriotism'.

There seems no reason to generalize further on the motivations and the abominations perpetrated by dictatorships, but one could just mention what occurred at Auschwitz under a recent dictatorship, where those who committed inhuman cruelties on the inmates justified these acts in court by explaining that 'they were just carrying out orders'. It does not seem much of an exaggeration to suggest that totalitarian regimes are simply applying the principles learned from the Anunnaki by ensuring that we effectively obey the commands of our superiors!

One confusing aspect of the genetic modifications by the Anunnaki to our native DNA just mentioned, was the quite separate hybridisation that resulted from matings between Anunnaki males and the 'upgraded' human females. According to Sitchin, this led to the birth of the cone-head giants or Nephilim. What may have been another form of genetic experimentation discussed later, is suggested by preliminary DNA analysis of the remains of the supposed Anunnaki-human hybrids discovered in Peru [60]. Here, the sexual roles may have been reversed in the two species in the production of early hybrids.

Figure 9. Image of a clay tablet showing Enki and Ninhursag (centre and right) in a laboratory while creating the perfect human being (sketched from a figure in Evans[13]).

Construction of an Ark to Escape from the Post–Ice Age Flood

The Sumerian accounts of Sitchin identify Enki as the one who secretly provided detailed specifications for the materials and design needed by Utnapishtim (the Sumerian equivalent of Noah) to construct the Ark. This itself argues for Enki's material existence. Favoured humans floated in the Ark and survived the Great Flood. Enlil, who knew in advance of the onset of the Flood, had avoided informing his human servants of it when he secretly moved the Anunnaki up to their space station. He apparently saw this as an opportunity to rid the world of his now inconvenient and rapidly multiplying human vassals. However the Ark rescued a small group of humans, as well as the 'seed' of their important livestock and plants. The similarity of this myth to the biblical story of Noah's Ark is evident. The Great Flood was dated by Sitchin at 11,000 years ago, while the rebuilding of cities destroyed by it was reported to have taken place about 7,000–8,000 years ago (see Annex table). In the meantime, it is presumed that the hybrid Anunnaki-humans, here referred to by the biblical terminology as the Nephilim, took advantage of the temporarily low numerical abundance of humans to erect regional Megalithic civilizations, which may at a certain point have descended into warfare between themselves over control of their respective regions.

Colonization of Southern Africa to Mine Gold, and Our Later Migration North

According to Sitchin, the city where the Annunaki first reigned was Eridu, the oldest Sumerian city in south-eastern Iraq. It was the home of Enki, who later became one of the gods of the Babylonians. The name Eridu is the basis for the word English-speaking people use for the planet Earth, and the region referred to as E.din is translated as 'Home of the Righteous Ones'.[1]

In southern Africa, one of the most mystifying and extensive series of ruins discovered near Maputo, Mozambique, was suggested by Michael Tellinger[11] to date back to roughly 200,000 years ago. The many ruins found there were presumed to be accommodation for the humans used to mine for gold, which may have been an operation involving slavery. As noted earlier, the upgrading of the proto-humans as a workforce happened on the suggestion of Enki, after Anunnaki workers refused to work in the dirty and dangerous goldmines in southern Africa (Evans[13]). Tellinger describes archaeological discoveries by a team of leading scientists studying this ancient southern African city. They seem to confirm that the Egyptians inherited some of their technical knowledge from this earlier civilization that existed in southern Africa more than 200,000 years ago.[11] Of relevance to this hypothesis was the discovery there of proto-Egyptian statues in dolerite rock: notably, a bird resembling Horus, a model of a sphinx, and inscriptions of the Egyptian archaic symbol, the Ankh.

Proto-Sumerian symbols were also found, included a petroglyph of a winged disk and carvings of Sumerian crosses. These finds led Tellinger to suggest that this unnamed city was a predecessor of the Egyptian civilization that began thousands of years later (although perhaps the Sumerian symbols were brought to Southern Africa by their 'captors'?). It was believed that eventually any escapees would have learned once again to live in the wild and create their own mythologies from their past experiences, such as the widespread subsequent human belief that gold belongs to God and not to human beings. Eventually it is supposed that by at least 10,000 BC, they had migrated northwards to the Nile valley, providing a human contribution to the early Egyptian civilization. Local traditions of the native peoples of Southern Africa say that here was where humanity was created by the god known in the Zulu language as 'Enkai'—evidently the same deity as Enki in Sumerian inscriptions.[11] Further evidence of this distant African cultural connection to Sumeria is the word 'Abantu' used to describe black South Africans, which could simply mean the 'children of Antu', the Sumerian goddess of the same name.

The vast area occupied by some million circular stone ruins in Southern Africa was described and a provisional age of some 285,000 years assigned to these ancient constructions.[11] Tellinger, in *African Temples of the Anunnaki: The Lost Technologies of the Goldmines of Enki*, provides evidence for this previously unknown civilization in southern Africa, which preceded the Egypt of the pharaohs by more than 150,000 years. If we are to assume that these ancient constructions were dwellings, Tellinger suggests they housed a population of at least 10 million people. The purpose of these constructions is not yet resolved, but since early explorers found many ancient mine shafts close to the ruins, gold mining was the obvious occupation of their tenants. The question then is: Which civilization, if not the Anunnaki, was importing large quantities of gold at this early date? The explanation proposed on the clay tablets was that the gold extracted was sent back to their home planet to be added to the atmosphere of Nibiru in monatomic form, to reduce a growing heat loss from the planet—a problem which seemed to have had critical importance as a climate crisis for the Anunnaki.

There is another unresolved question: How could these circular ruins, with no evidence of door openings, be dwellings? Their location fits in with Sumerian descriptions of gold mining initiated by the Anunnaki in Africa, but the closed nature of these buildings suggests their labour force was retained within. If so, might the early human relationship with their masters, the Anunnaki, have been a form of slavery rather than cooperation? There also seems some evidence for an underlying technical basis for these circular structures which appears to fall outside the modern theory of science, but is perhaps is linked to concepts of sound transmission and magnetism? [11] One suggestion is that their designers, the Anunnaki, used the stone circles as amplifiers of terrestrial sound resonance for energy generation? A type of resonating hand-held stone cylinder found there has been speculated to have acted as a linkage to the possible use of sound energy for moving large blocks of stone.

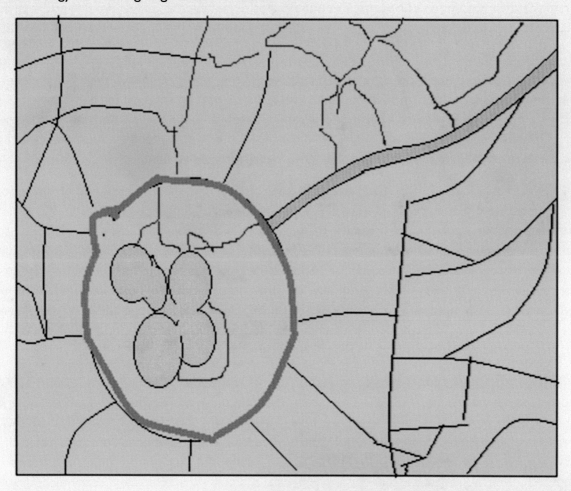

Figure 10: A southern African example of the ancient dwellings drawn from ref 11, with a wall constructed around each of them. There is also a wall enclosing the 'road' leading to it (but no doorway is evident to the exterior or to the circular 'rooms' within the outer wall). These structures were originally considered to be cattle kraals, but the linked pathway from the upper right corner shows no obvious entrance to the walled compound!

Conflicts between the Anunnaki and Their Followers

As mentioned on the Sumerian tablets, seen from a certain perspective, Enki may be considered mankind's greatest benefactor, since it was he who suggested that the Anunnaki create modern humans from the pre-hominids then present on our planet. From the opposite perspective, many of our negative qualities may have come from imitating our Tutors. Nonetheless, another of Enki's positive actions was that he later rescued the equivalent of Noah, our ancestor, from the Flood.

By going on strike, the Anunnaki workmen had interrupted the supply of gold to their home planet and alarmed Anu the ruler of the planet Nibiru. Anu came to Earth with his son Enlil the half-brother of Enki, to resolve the issue. In a measure designed to resolve antagonisms, the two half-brothers were told to draw lots to determine their terrestrial areas of influence. All areas north of the Mediterranean and Indian seas were assigned to Enlil, but Africa and the oceans were assigned to Enki.[1]

After fifty years of practice, Zecharia Sitchin had become an expert in the interpretation of Sumerian clay tablets and concluded that humans had learned key aspects of our current mode of life from the Anunnaki. Included here are our adaptations to urban life, the techniques of extensive agriculture and irrigation used at this early date, and later, a tendency to inter-urban aggressions. The early Near East emerged as a horticultural laboratory, producing edible seeds, flax, oils, and a variety of fruits. It is hard to imagine a newly agricultural society could produce such items without considerable previous experience in horticulture. In the highlands of the Near East, the domestication of animals also began, and this would have been difficult before mankind acquired the habit of living in fixed social structures of some size, which permits specialization of livelihoods such as animal breeding separately from migratory hunter-gathering of food.

The Middle Stone Age was often referred to as the 'Age of Domestication' lasting from 11,000-10,000 BC, and pottery appeared around 7,000 BC. The ancient Mesopotamian and Egyptian civilizations arose in an astonishingly short period, by what some experts have called 'stunning abruptness'. Apparently, both societies immediately acquired almost all the features subsequently recognized as characteristic of sophisticated city life. The startling abruptness noted for the early Pharaonic civilization in the passage from a Neolithic tribal society to a monarchy with monumental architecture, writing arts, and crafts characterized a 'ready-made' luxurious civilization. All this apparently without early development stages (Walter Emery, quoted in [4]).

According to Tellinger, the genes we received from the Anunnaki, in addition to those contributing to domesticated behaviour, had an endemic tendency resembling that of the Anunnaki, to use aggression and warfare to solve problems. In other words, this tendency was culturally transmitted to their human servants. Their tools of warfare were sophisticated, and from his interpreted accounts, Anunnaki conflicts between followers of the two half-brothers even used atomic weapons brought from Nibiru. In one case, estimated to have occurred in 2,024 BC, these weapons were aimed at a spaceport located in the Middle East and at several human cities in Mesopotamia. An excerpt from a Sumerian inscription describes the result of this conflict,[11] whose features can be recognized from more recent human experience with nuclear warfare in the following quotation, where the 'gods' are evidently the Anunnaki:

> 'On all the lands, from west to east, a disruptive hand of terror was placed. The gods, in their cities, were helpless as men! A death-dealing wind born in the west its way to the east has made, its course set by fate. Poisoned weapons in the great plain then unleashed. That an Evil Wind shall follow the brilliance we knew not! They now cry in agony. In their holy cities, the gods stood disbelieving as the Evil Wind toward Shumer made its way. Escape to the open steppe! To the people I gave instructions; With Ninki, my spouse, the city I abandoned. The Evil Wind against Nippur was onrushing. In his celestial boat, Enlil and his spouse hurriedly....' [90]

According to Sitchin's translation, human refugees were advised not to look back while leaving the city of Sodom; not for fear of being converted into stone but to avoid the effects of radiation following the nuclear bombardment. The following excerpt is a similar description from an ancient Indian text, the Mahabharata[4] (6,500 BC at the latest), which also features aspects of what can be identified as nuclear warfare which took place in ancient times:

> 'The cloud of smoke
> rising after its first explosion
> formed into expanding round circles
> like the opening of giant parasols ...
> It was an unknown weapon,
> An iron thunderbolt,
> A gigantic messenger of death,
> Which reduced to ashes
> The entire race of the Vrishnis and the Andhakas.
> The corpses were so burned
> As to be unrecognisable.
> The hair and nails fell out
> Pottery broke without apparent cause,
> And the birds turned white'.

John F. Caddy Ph.D.

Judging from fairly detailed accounts summarised on [53,99], especially where there seems extensive desertification, extensive evidence of atomic explosions in the distant past has been found. In one case, a layer of radioactive ash covers a three-square mile area in Rajasthan, India. Scientists were investigating a site where a housing development was being built, and it was reported that a high rate of birth defects and cancer had occurred in the area under construction. Scientists unearthed an ancient city where there was evidence of an atomic blast dating back to between 8,000 and 12,000 years ago (Fig 11).

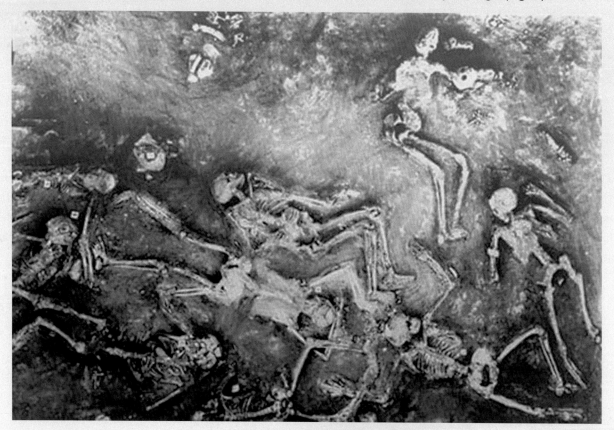

Fig 11. Painting of the skeletons found at the Mohenjo Daro site of an ancient atomic explosion (source: Wikimedia[98]).

Nonetheless, apart from nuclear warfare, the key catastrophe that effectively wiped out most civilizations and their historical records before 12,000–13,000 years ago, was a global mega-flood. Some experts believe this was related to a geological/cosmological trauma experienced by our planet, possibly as a result of the passage of Nibiru within the solar system. Sitchin's work, and others subsequently, inspired me to look for new information that has since come to light, suggesting a possible associated schema for early human evolution. This framework is intended for criticism and invites other suggestions for an alternative prehistorical framework, if one is available. An integrated multidisciplinary approach seems to be the only way we can make progress in understanding key events in our distant past where technological discoveries formed a part.

Unimaginably Long Reigns: A Function of Planetary Stability?

The royal canon displayed at the museum of Egyptology of Turin dates back to Ramses II, and presents a list of all the pharaohs who supposedly reigned in Egypt and the durations of their reigns. This list includes not only the historical pharaohs, but also the 'divine pharaohs who came from elsewhere' and reigned for 13,420 years before the recognized first dynasty of Menes. By consensus, information on the reigns of the later pharaohs on this list are accepted in official circles, but the early lineages are regarded as mythical—probably because of the improbably long reigns of these early pharaohs.

All available sources speak of a lineage of gods who each reigned for several hundreds of years, for a total of 23,200 years. They then speak of the 'Followers of Horus', who reigned after them for 13,400 years. After that came the names of the 'normal' sequence of Pharaohs we know from hieroglyphic inscriptions and Egyptian monuments. That early divine and semi-divine sovereigns might each have reigned for hundreds of years is unacceptable to Egyptology experts, just as it was in the case of the early Anannuki kings in Sumer. At the same time, we should ask ourselves why we seem to accept the several hundreds of years of lifespans for prophets of the Bible such as Enoch, who is reported to have lived for over 360 years. Some studies speak of the many factors which could explain a much slower aging process for the Anunnaki through their greater scientific knowledge, or resulting from a change to terrestrial gravity. Arthur Mendez, co-founder of the Church of God in South Texas expressed the view that the rate of decline in longevity from pre-flood times recorded in ancient texts, matches the rate of decay observed in organisms when they are exposed to radiation or toxins. The two main factors that should be borne in mind however, are (1) the environmental conditions that might have led to the evolution of extremely long lifespans on the Anunnaki planet of origin, and (2) their advanced knowledge of the genetics of the Anunnaki and later, of human beings.

According to Sumerian accounts, Anunnaki lives were very much longer than for humans, but their life expectancy apparently was somewhat shortened while resident on Earth. If we follow the duration of reigns of the Anunnaki or their hybrids in the Sumerian Kings Book, their kings appear to have been close to immortal, with lifespans mentioned of tens of thousands of years. Explaining such a very long lifespan is difficult without any clinical evidence[24] but if we were trying to explain such an anomaly (when viewed from a human perspective), it might be associated, I suppose, with the infrequent reproductive activities of a species which evolved on a planet without a seasonal cycle: given that this planet and its associated dwarf star, travels for many centuries beyond the range of the sun's rays? A more specific problem arises from our natural doubts on lifespans in excess of 10,000 years, as in several cases mentioned. However speculation on the possibilities, on the one hand of clonation as a source of new stem cells, or even induced hibernation as needed for long space voyages, cannot be excluded as mechanisms. The doubts that arise may have a practical consequence, since this information was used to count back to find the date of arrival of the Anunnaki on our planet. Could their arrival have occurred much later than 400,000 years ago if the ages of early monarchs were exaggerated?

A slow metabolism, a long life and low fecundity, could nonetheless be postulated as the reasons why their continuity on Earth seems to have eventually terminated. The disgust of Enlil, chief of the Nephilim on Earth at our frequent reproductive activities would be understandable since the early onset of sexual activity in humans, from the perspective of a 'millennial' species, would be seen to long precede the age of social responsibility. This is hypothesized as one of the reasons why Enlil abandoned the humans on the ground at the time of the Great Flood. The other reason he frowned on this human behaviour was the enthusiasm of young Anunnaki males to mate with 'upgraded' female humans, interspecies sex being forbidden by Annunaki laws. These rules must also have been a feature of a monarchy where a king with a millennial lifespan was assigned divine status over lower castes, in a species presumably subject to rigid reproductive rules dictated by their long lives.

Interestingly enough, the lifespans of the longest living vertebrates on Earth are the Greenland sharks, which live as long as 300 years and mature at age 150. Bowhead whales often live for up to two centuries if not hunted. Both vertebrate species mentioned inhabit waters close to the freezing point year round, and at low-light levels. Bacteria located several metres below the mud surface in cold, anoxic sediments have been alive but in suspended animation for millions of years. If we consider that life on Nibiru, which must be a cold planet for most of its orbit, might have been spent during the early evolution of the Anunnaki with limited natural food resources. Then a long lifespan, possibly with hibernation and a very limited rate of reproduction, might be called for? Frankly, further speculation on this question is pointless with currently a lack of information on the lifespans of ET's. It could be useful, however, to avoid contrasting their long lifespans with our own short lives. In our case, a single generation has been estimated to range between twenty-five and thirty years, and we may reasonably consider that our own lifespan is rather short. As an example, the working life for a modern human being usually begins around age eighteen to twenty in developed countries, followed by a working life rarely exceeding forty years before a short period of retirement. This is nonetheless a favourable situation compared with the much shorter lives prevailing a few centuries ago in developed countries, and still nowadays in parts of the developing world.

An article in an Italian newspaper today gave an example of the speed of current technological development in today's world and how it is implemented by progressively younger technocrats. A name mentioned was Elon Musk, a South African by birth, who earned his fortune through creative activities in different US industries before deciding in 2002 to found 'SpaceX': a company almost alone in the private sector in acquiring space-going experience, which planned to land a spaceship on Mars within a decade or so. A 31 year old engineer who joined a team of SpaceX engineers found himself almost the oldest in the team where technical decisions were made by 20 year olds, then reviewed by the 'older' thirty-year old engineers. Does this mean that human genius occurs mostly early in life? And for long-lived intelligent organisms, what would be their age when important discoveries were made?

Another example in today's newspaper described how a 16 year old student manifested every Friday in front of her Parliament, demanding immediate action from adults to correct the seriously deteriorating climate that was destroying the prospects for her generation. News stories on her actions led to similar absences from school world-wide and the largest political manifestation by adolescents seen so far, expressing the reasonable view that a lack of effective action by adults to counter climate change would principally destroy the prospects of her own generation.

According to an article in the *New Scientist* (22 September 2018) entitled, 'Only One in Five would like to Be Immortal', more people are worried now about the implications for overpopulation from radical life extension than are optimistic about it, and 44 per cent even felt that we should accept our natural lifespan. Nonetheless, some gerontologists believe that a radical life extension may be available to people alive today. One positive aspect of a relatively short lifespan for intelligent species is the continual renewal of intelligent minds in the population, given that the occurrence of genius in human beings often manifests early in life. This means that (unless constrained socially) their inventions can result in rapid and radical changes to the current technological and social situation such as we have witnessed in recent decades. Even a cursory inspection of our recent history should convince you that a society with a high proportion of young well-trained experts can be competitive with a society where older experts are dependent on earlier theoretical and practical discoveries.

Nonetheless, our relatively short lifespan seems linked to diurnal and seasonal stresses imposed by a rapidly rotating planet, and the high levels of radiation and chemical contamination our planet has received over recent millennia. These factors have perhaps acted negatively on our physiology to keep us at best centenarians. Or is our short lifespan and high fertility a design feature planned for a servant race? Certainly, the age at sexual maturity of terrestrial mammals and other vertebrates is short—rarely much in excess of five years, confirming the stresses terrestrial mammals are subject to from predation and environmental impacts. Life histories studies on marine organisms show that their group mortality, and hence life expectancy, is dependent on the number of risks an age group is subject to in a unit time. It could be possible therefore, that on planets where environmental conditions are relatively constant, lifespans could be longer, while species inhabiting a planet with frequent perturbations involving high risks to life, would need to reproduce earlier and would have shorter lifespans.

A sociological consequence of our short lifespans that is of serious concern is whether a working lifespan of some forty years will allow government officials to accumulate adequate experience to make complex decisions on the future quality of the terrestrial ecosystem? This question illustrates a critical situation we are now facing: will governments take into account close to cyclic events linked to our planet's rotation cycle of close to 25,000 years? Turning the decision over to an effectively immortal sentient computer may not be a wise idea, but perhaps could be preferable to human decision-making, given that mythology is the substitute for historical memory in short-lived intelligent species. This decision would however conflict with one of the strong points of our species; the ability of brilliant youngsters to overturn existing paradigms!

A tendency towards immortality was just mentioned for the Anunnaki, but the book of Genesis also attributes slightly less extreme longevity to the immediate descendants of Adam after the Great Flood. Lifetimes of humans were reported to extend from 365 years for Enoch to 969 years for Methuselah, and this makes one wonder as to the real genetic provenance of these personages, since these age estimates (possibly somewhat exaggerated?) approach the chronologies Sumerian tablets provide for the even greater ages reached by Anunnaki kings and early pharaohs. Might an intervention using Anunnaki genes explain the longevity of these ancient biblical personages as recorded on the Sumerian tablets? The Old Testament of the Bible, written some thousands of years following the Flood, recorded a progressive decline in the lifespan of the human patriarchs—from Noah, who lived to be 500 years old, until Abraham who reached only 175.[24] Could the decline have been a result of the radioactivity that must have been diffused in the Middle East from the nuclear wars also described on the Sumerian tablets and mentioned later?

Although the reproductive procedure for 'improved' humans was not described in detail, the fertilization and modification of a *Homo erectus* ovum was presumably in vitro, considering the glass vessels held by a 'technician' in Fig 5.

The tablets recounting this procedure were written by scribes with limited scientific knowledge, but describe in simple terms what may have been clinical trials with pre-hominid eggs, fertilized by Anunnaki males, then implanted into a 'production line' of 'birth goddesses' consisting of Anunnaki females. Given that the skulls found in the Maltese Hypogeum were cone-headed, is it possible that the statues of enormously fat women[38] found in ancient ruins there, represented these Anunnaki birth goddesses?

Sitchin wrote that whatever method the Nephilim used to infuse genetic material into the biological makeup of the hominids they selected, the end result with this method was quite probably a sterile offspring, at least initially. The pre-hominids or 'black heads' as they were referred to, were moved from Africa to E.Din in Mesopotamia and further engineered to increase their breeding capacity and presumably also their fertility, given the high demand for their services. The upgraded human females resulting were apparently also fertile with Anunnaki males, who began to mate with them, and giant hybrid hominids or 'coneheads' resulted, who eventually formed the ruling class for megalithic societies, and may have been responsible for the frequent wars between their cities which resulted.

Evans quoted Sitchin's interpretation of the Sumerian clay tablets, in particular his description of the reaction of 'The Watchers', as the Anunnaki personnel stationed off-planet were called. These were confined for long periods on space stations above Earth, or on Mars—stationed there to transfer loads and guide spaceships going to and from Nibiru. The Watchers were on Mars for thousands of years but did not find life there congenial. Once they saw images of the 'improved' female humans, they rebelled, came down to Earth, and took them as mates. A similar sequence of events is later expressed by the book of Enoch in spiritual terms: This recorded that in olden times, two hundred heavenly 'angels' as Enoch referred to them, rebelled against God in heaven. They came down to Earth on Mount Hermon searching for the upgraded human females, 'and defiled themselves with them'. This resulted in the birth of 'great giants' who eventually 'consumed all the acquisitions of men'. The book of Enoch[39] also noted that these activities provoked God to arrange that the cursed giants waged war against one another 'so that they may destroy each other in battle'. This final consequence of hybridization perhaps reflects earlier events in our prehistory: notably wars between the war-like Anunnaki or the cone-head giants in charge of competing megalithic societies, involving human vassal armies.

The urgency to mate with the genetically modified human women expressed by the Watchers off planet may have been due to prevailing restrictions on the mating of lower-caste Anunnaki on Nibiru, which could have been a requirement designed to prevent overpopulation of their planet, given that they are a very long-lived species. This cross-fertilization between two species was reported on Sumerian tablets to have resulted in hybrid giants, the so-called Nephilim, who in the Bible were called the offspring of the 'Sons of God' and the 'Daughters of Men', and they may perhaps be identified with the 'Cone-heads' mentioned earlier. Thus once again, the earliest biblical records recording spiritual events seem in some cases to reflect real events that occurred much earlier in history. The hybrid offspring, often referred to as 'demi-gods', were long-lived, and warfare between factions ruled by Anunnaki-human hybrids were described in Sumerian inscriptions. The spiritual actions described by Enoch thus correspond well with the actuality recounted in the Sumerian tablets, including the later battles between lineages of Anunnaki, their hybrids, and their human troops.

Other Humanoids in Our Planet's History: The Megalithic Civilizations

Several articles on the Internet discuss a past court action that investigated the possible involvement roughly a century ago of an eminent institution of natural history in a historical cover-up of the tens of thousands of human remains between 6 and 12 feet in height that were excavated from burial mounds all across America. The skeletal remains of giants were often portrayed photographically in local newspapers (e.g. Fig 12), and some of the skeletons were reported to have been sent to the museum. On later investigation, these were not to be found in the museum, and supposedly were either reburied or simply disappeared. The high-level museum administrators were believed to have tried to protect the accepted mainstream chronology of human evolution at the time, which contended, and still contends, that America was first colonized by Asian peoples migrating through the Bering Strait 15,000 years ago. There were thousands of giant burial mounds all over America, which the native nations claim were there when they arrived. These contained traces of a civilization reportedly involving the complex use of metal alloys.

Fig 12. Skeleton of an 8-ft giant compared with a 6-ft man [100] **(from The Steelville Ledger, June 11, 1933).**

As already mentioned, a distinctive humanoid form with an unusual cranial structure was reported from the Megalithic era: giant hominids upwards of 2 metres in height, with red hair and often with elongated skulls.[31,41,60] Their remains have been found in Eurasia and South and North America, and a skull with similar features has recently been located near Arkaim, in central Russia. This last individual dated from the seventeenth century BC at a site referred to as the 'Russian version of Stonehenge'.[41]

In *Megalithic Origins: Ancient Connections between Göbekli Tepe and Peru*, Graham Hancock[42] also observed that in the Zagros region of Iraq twenty-seven cranially elongated skulls were discovered from around nine thousand years ago. These were either deformed intentionally or, in light of the hypothesis presented here, could have been Anunnaki or their hybrids, originally from Sumer.[1] Hancock noted that their remains closely resembled an elongated skull discovered at Kilisik, near Göbekli Tepe. Long skulls have also been unearthed at megalithic sites in Peru and Bolivia from different dates. Hancock observed that a surprising number of such skulls have been found worldwide near megalithic sites in Egypt, Micronesia, North America, Ukraine, France, Austria, and Malta. Numerous small statues found in Iraq from around 6,500 BC also depicted thin-faced humans with very long skulls. Similarly elongated skulls have been found at Machu Picchu, Cuzco, in Ecuador, Honduras, Chile, Mexico, and Colombia. The subsequent cultural preference for creating children with elongated skulls in human tribal peoples, has been found on almost every continent on Earth, suggesting a cultural connection of great antiquity.[77] Why were long skulls so prized? Fig 5 shows that high headgear, falsely implying an impressive vertical extension of the braincase has often been used quite recently by those in high positions in society.

The pertinent question is whether this cranial deformity, called dolichocephaly, exists naturally? Was the significance of its persistent emulation as a desirable characteristic by so many people, a reflection of the true appearance of those once holding high positions in society? There seems an association here with how the gods were viewed in early human attitudes to this condition: an elongated skull presumably was a symbol of status, high rank, or wisdom. Could it be that the use of tall headgear by archbishops, military personnel and wealthy business men perhaps also reflects a distant memory of this symbolism? (Fig 5)

Hippocrates was the first to suggest that cranial elongation might in some cases be inherited. With true dolichocephaly, if there is an increase in brain capacity it would be reflected in the genotype. Dr Tschudi, an expert on Peruvian archaeology, is quoted as pointing out that deciding between a natural inherited condition and a deliberate deformity imposed during childhood, rests on insufficient grounds if the observations are made solely on crania of adults. However, he reported that two mummified children (both scarcely a year old) belonging to the Aymaraes tribe, and a foetus found within a mummy of a pregnant woman, all show the same cranial elongation. These long skulls may have been the elite of society, and some researchers believe they had enhanced telekinetic abilities, related to how they 'moved' huge stones. On Easter Island, Hancock reported that there is a unique solo statue in the museum depicting a very odd looking female 'long-head' and that the legends there state that the Moai or stone statues 'moved themselves' or were 'hovered' into place by 'Manna'—the ancient Hawaiian equivalent of 'prana'. Returning to the main hypothesis of this study, is it possible then that these 'cone-headed' creatures were either descendants of the Anunnaki, or Anunnaki hybridization with humans? Ancient skeletal remains for the very tall 'cone-headed' humanoids[31,41] just mentioned (figure 7) show different patterns of cranial fissures from those typical of our species.

Similar elongated skulls were also found within the 6,000-year-old Hypogeum in Malta (Fig 13) when it was opened more or less a century ago. The Hypogeum in Malta is a Neolithic subterranean structure dating back to 3,300–3,000 BC. From the abundant skeletal remains found in it when it was first opened, it may have been constructed and used by 'cone-headed' humanoids (Fig 13).

Fig 13. A chamber of the Maltese underground temple the Hypogeum of Hal-Saflieni. (From Wikepedia).

The Hypogeum is an underground temple with unique acoustic properties and a central chamber that resonates at 150 Hz. It is not clear why it has such evident sonic properties, except that the acoustic sciences were apparently well developed in Megalithic civilizations and used for practical purposes. Its design and construction, and that of other megalithic structures already mentioned, suggest advanced acoustic applications. The technical crafts of precisely shaping and transporting enormous blocks of stone were in advance of our own, and there is no evidence of motorization.

'Amazing! Why hasn't the news covered these things before? How can they be all over the world and no one talks about it?' This comment by a reader of a website showing adults that were naturally dolichocephalic (see Fig 14) is the natural reaction of disbelief to the suggestion of the authorities that head-banding of babies was the cause of this 'malformation'. This appears to be the usual way of disguising the apparent ET influence on their origins. It however seems barely credible as a starting point for a discussion of all elongated skulls. Analysis shows that in the admittedly human practice, the cranial volume after head-banding of human babies remains basically the same, whereas the 'cone-headed' hominids had a much larger cranial capacity, as would be expected from a hybrid with a species much larger than humans, and shows other structural differences. A more reasonable suggestion for the strange head-banding tradition recorded around the world is that this was a form of imitation by the *Homo s. sapiens* labour castes of their megalithic masters, who were naturally built that way.

This issue points to a sociological problem following the discovery of the remains of ancient giants. But professional archaeologists and anthropologists seem to have remained strangely silent on these discoveries despite the existence of significant information sources and skeletal remains. According to *ancientaliens. com*, more than two hundred cone head skulls have been excavated from different locations worldwide, including skeletons of giant humanoids 2 to 4 metres in height, discovered in burial mounds in North America. Some of these North American giants (believe it or not) had a double row of teeth![51] Could these latter specimens have been early 'mistakes' in genetic processing that were then banished into the North American wilderness, or is this an Anunnaki dental characteristic? Their discovery seems to have been well documented a century ago in local US newspapers, but where the skeletons are now that were then sent to museums often remains a mystery. The restitution of skeletal remains to American tribes for reburial seems to have been the official response to enquiries as to their whereabouts, and in 1993, newly enacted US legislation, dubbed the Native American Grave Protection and Repatriation Act (NAGPRA), required that *all Native American remains collected by institutions or archaeologists on federal land using federal funds or permits be made available to the rightful owners or reinterred with the tribe of origin.* However, it seems debatable that the North American giants were necessarily related to current tribes of Native Americans.

According to the page 'Alien Cone Head Skulls' found on the website *ancientaliens.com*, there were only slight variations in size and shape between individual cone-headed skulls, which dispels the assumption that they were a rare deformity or mutation, or a result of deliberate cranial deformation. In his *Nephilim Chronicles: Fallen Angels in the Ohio Valley*,[49] Fritz Zimmerman included more than three hundred accounts of giant skeletons researched from ten thousand works in the Indiana University Archives which described remains recovered from more than two hundred prehistoric burial mounds in the United States. (For example, in the American Southwest, a tribe of red-haired, six-fingered giants was discovered mummified in a cave. Their height spanned up to 12 feet tall[40,49,51]).

Michael Cremo and Richard Thompson,[48] in their *Forbidden Archaeology*, a nine-hundred-page work, asserted that a cover-up of archaeological finds in North America occurred a century ago, although why this should have occurred is just as mysterious as the apparent cover-up of recent abundant sightings of UFO's. The reasonable conclusion as stated above, seems that these giants were either Anunnaki or hybrids of Anunnaki with humans. This hypothesis also provides a basis for the technical capabilities associated with megalithic construction elsewhere, requiring techniques which presumably were well within the knowledge base of the Anunnaki.

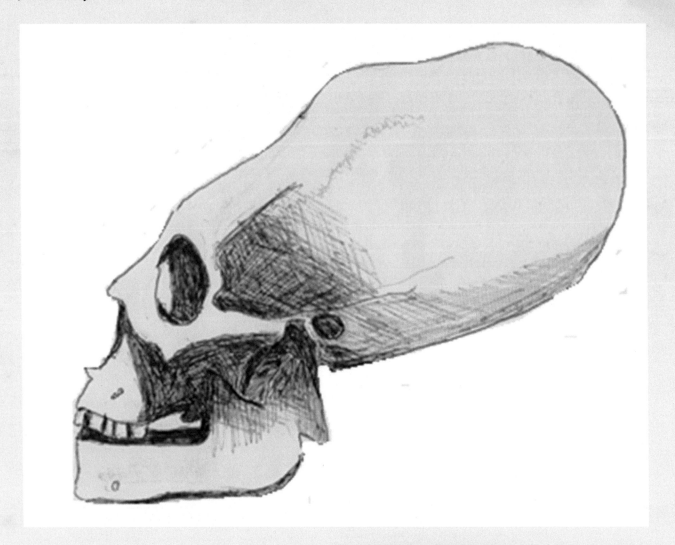

Figure 14. Sketch of a cone-head skull (drawn from a photograph in 'Alien cone head skulls', found on Ancientaliens.com).

In an attempt to explain the occurrence of Sumerian artefacts in other areas of the world, the idea expressed by some commentators was that these Anunnaki or their hybrids may have made the long sea journey from the Black Sea region to South America. This ignores the fact that advanced civilizations of Anunnaki origin before the last Ice Age may have had modern means of air and sea travel. (See, for example, engravings of aeroplanes, ships, and submarines in the ancient Egyptian temple of Seti [61], a mural in another Egyptian temple showing the launching of a rocket from an underground silo [91], and pre-Colombian American gold broaches[62] in the form of aircraft). It is not difficult to imagine that small groups of human colonists could have been transported by modern vehicles constructed by a space-going species to locations they had identified from space as potential sites for gold mining, or to investigate the possibility of Anunnaki colonization of Earth.

Presumably colonization also occurred in North America where, as mentioned, nineteenth-century farmers digging into commonly occurring burial mounds in the State of Virginia,[63] found skeletons of individuals reaching 2 to 4 metres in height (Fig 12). Again, the offspring of these giants appear not to have survived to contemporary times, but why not? Battles with human tribes who were capable of rapidly multiplying their populations may be one possible explanation. The conquest of the giant Goliath by small, but accurate David, shows that greater size is not always an advantage. Low fertility or quasi-sterility of this type of hybrid hominid may have been a contributing factor. A short generational time is a competitive feature for human beings when encountering a long-lived species, although our rapid rate of population growth and high fecundity are now of course the basic reason for the many international crises we now face as a result of overpopulation and competition for limited resources.

Before leaving the subject of genetic modification, we should mention a recent discovery. On the desert peninsula of Paracas along the Peruvian south coast, multiple cone-headed skulls were found in a graveyard dating back some three thousand years.[52] These skulls had a 25 per cent greater capacity and weighed 60 per cent more than an average human skull. DNA analysis was carried out on bone samples from them by three institutes in America and Europe. Their nuclear DNA was found to resemble that from ancient skulls found in the Caspian and Black Seas regions where similar cone-headed humanoids have been discovered,[41] together with another example recently found in Antarctica [92]. (This discovery has, of course, been contested, as with most recent surprising discoveries). The fact that the Paracan nuclear DNA shows no similarity to native peoples in South America is significant, but the key point to note was the characterization of the mitochondrial DNA: this was found to be very different from that of either modern humans or other terrestrial animals.[82] It raised an obvious question given that mitochondrial DNA is associated with the maternal cell: was the genome of the ancestors of these ancient Paracans the result of genetic manipulation with ET genes?,

One possible mechanism one could consider here, that might perhaps have given this type of result, was the injection of a fertilized nucleus of a human from Asia or the Middle East into the ovum of an Anunnaki female. Would the resulting offspring differ from the result of cross-breeding between Anunnaki males and modified human females, also discussed here? In this hypothesized case, would the offspring also have differed from the results of gene splicing suggested by Sitchin to have resulted in modern humans? Many questions remain to be answered on this and other subjects raised in this review, and further DNA analysis seems called for.

I ask again: Why did this community of cone-head 'hybrids' not apparently survive long after the period of megalithic construction? Was this a result of reduced fertility, or was it caused by warfare with adjacent human communities? Did the transition from megalithic societies to those run by less advanced humans occur despite the possibility that nuclear weapons were employed by Anunnaki hybrids to exterminate humans and their cities? Or perhaps the Anunnaki and their hybrids left their cities for more favourable environments elsewhere such as Antarctica and Siberia before their climates deteriorated? It would also be interesting to know more about the role of our human ancestors in megalithic communities ruled over by the Anunnaki and their hybrids, but nothing remains concerning the sociology of these hypothesized mixed species communities.

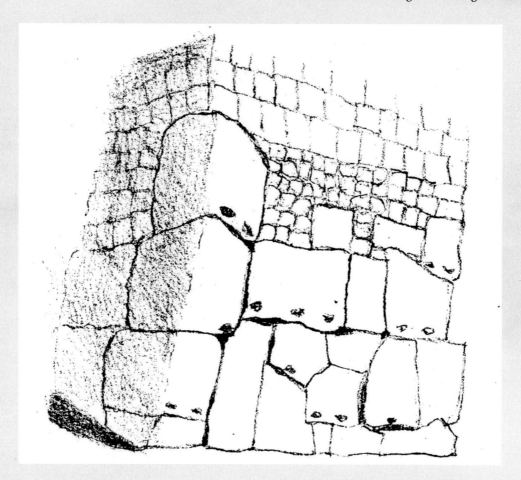

Figure 15. Configuration of a megalithic wall at Cusco, Peru. The photographer D. Olsen [101] remarked on the contrast between the precision megalithic stonework below, and the rough configuration of the smaller rock wall above which was added by the Incas in their reconstruction of the ruined megalithic constructions left by earlier occupants.

Megalithic Constructions

Now consider the major problem that megalithic civilizations pose for archaeologists who are not advised by publication outlets to refer to extraterrestrials in their theses. This profession has to explain how stone or copper tools enabled 'primitive man' to move and shape blocks of granite and other hard rocks, weighing hundreds or thousands of tons from distant mines and assemble them dressed perfectly as monuments at a variety of locations in Africa, Europe, Asia, and South America. This capability can only be rationally explained at this early time by supposing that extraterrestrial technology or some form of telekinesis was employed. It would not be surprising if, after the age of the Anunnaki, their survivors (i.e., the Anunnaki-human hybrids or their trained human servants), still had access to their ancient construction techniques. If this were the case, it would explain the diffusion of buildings made with giant blocks of hard stone in Egypt and at other sites around the world, such as the huge megaliths in the Baalbek site, Lebanon[65]. The even larger blocks of hard stone used to construct mysterious edifices in Siberia are surrounded by an even greater mystery [88]. The Baalbek citadel (Fig 16) they say, fell into ruins at the time of the deluge, and was rebuilt much later by a race of giants under the command of Nimrod. The structure of buildings in Cusco (Fig 15) also points clearly to megalithic inhabitants of a different origin before the Incas arrived.

Fig 16. Photograph from the 1890s of the Baalbeck temple showing the huge megalithic foundations of the Temple of Jupiter [96] (public domain).

Much later this platform of massive stones was used as the base for a Roman temple. Similar precisely shaped giant stone blocks were also used in megalithic constructions at Tiahuanaco and Cusco in South

America and in Egypt. The fact remains that hard rocks such as granite, cannot be shaped using the stone or bronze implements which were supposedly all that were then available. A precision of angles and inserts in megalithic structures is even less explainable, while their transportation without anti-gravity also remains a mystery. The suggestion made elsewhere that megaliths were created on site by some unknown form of cementification process cannot apply in this particular case where giant rocks were extracted entire from quarries.

A modern clue to the construction of megalithic structures may have been provided by a Latvian immigrant Edward Leedskalnin, who between the 1920s and 1950s singlehandedly built one of the greatest wonders of architecture in modern times: Coral Castle, Florida: consisting of massive blocks of coral rock weighing hundreds of tons.[81] Leedskalnin refused to share his construction secret, but claimed he built the castle with nothing more than an ancient understanding of magnetism, before moving the temple singlehandedly to another location—all of which defies our modern-day understanding of physics.

Let us return to the Megalithic Age and the huge structure at Baalbeck (Fig 16). This was hypothesized to have originally been a spaceport, with a temple built onto it much later, and later still another temple by the Romans. The megalithic walls of Cusco, and what may also may have been a spaceport at Tiahuanaco, Bolivia are notable also, before some unspecified mega-disaster disassembled what appeared to be complex buildings constructed from machine-cut hard rock identical components, or from an unknown form of cement. Some of these components show precisely measured and hollowed out interior spaces, which it would be difficult to replicate even with contemporary machinery. These are all structures that call for technologies far in advance of those that were available to Stone Age or Bronze Age civilizations, or to the Romans (or to ourselves); given the thousands of ton weights of blocks used in the foundations of the Baalbeck temple. Pyramids have also been identified in the Balkans, in South and North America, Asia and Antarctica, suggesting that the Megalithic Civilization was worldwide in extent before the Great Flood and that humans participated in it, presumably at predominantly lower social levels. Could this be a reasonable explanation for the tradition of head-banding of infants in some native communities—to seek a higher status for human offspring in megalithic societies dominated by cone-headed hybrids?

Notable among the Megalithic constructions are pyramids, the pyramids, which as mentioned are located worldwide, although those in Egypt are still in the top category of the marvels of the ancient world. They also form an unsolved mystery, not only as to who, when, and how they were built, but for what purpose? Although our science has not provided much help with these questions, hypotheses have been put forward suggesting that the Giza pyramid was a giant mechanism for generating pulses of energy, or was it used as a point of reference for incoming Anunnaki spacecraft.[1]

Sumerian verses quoted by Sitchin1 suggested the role the Anunnaki had in its construction:

> *'On the flatland, above the river's valley, Ningishzidda a scale model built,*
> *The rising angles and four smooth sides with it he perfected.*
> *Next to it a larger peak he placed, its sides to Earth's four corners he set,*
> *By the Anunnaki, with their tools of power, were its stones cut and erected.*
> *Beside it, in a precise location, the peak that was its twin he placed.'*

The religious distinction made by ancient Egyptians between the Pharaohs and ordinary citizens could reasonably be explained by the suggestion that at least at first, the Anunnaki or their hybrids reserved the top positions in society for themselves. This might also explain the elaborate mythology of rebirth or reincarnation described in Pharaonic mythology. It was apparently reserved for them - perhaps to regain by reincarnation their original place on the planet Nibiru?

The apparent linkage of the 'cone-headed', red-haired giants to ancient megalithic site where these relics were found is shown by the eight-thousand-year-old Maltese underground temple called the Ħal Saflieni Hypogeum. This suggests that these technically efficient Anunnaki or their hybrid Anunnaki-human descendants, are essential to our understanding of megalithic architecture. Perhaps the methods of transportation and construction they used were advanced developments in sonic and/or antigravity technologies originally discovered by the Anunnaki? Or perhaps their larger brain capacity incorporated teletransportation skills based on advanced pranic applications? Also, at various megalithic sites around the world a similar technique was used for holding large blocks in place, namely to pour liquid metal into I-shaped moulds linking together adjacent blocks. This implies a common methodology and origin, and perhaps a linkage exists between these construction techniques and the advanced lifespans of the Anunnaki. Not wishing to see their buildings collapse during their long lifespan, they took the trouble to construct them with giant irregular but close-fitting blocks that would be less subject to extensive cracking and would certainly have withstood earthquakes far better than human structures with small rectangular bricks and mortar.

A similar process was documented during the construction of edifices close to the Great Pyramid of Giza. According to a video made by the Farsight Institute in 2014,[64] Distant Viewers mentally visited the Giza pyramid during its construction and saw what appeared to be cloned hominids employed in the pre-Pharaonic civilization as a labour force under the control of an extraterrestrial supervisor. They precisely cut huge blocks of granite and other hard rocks by what can only be called machine tools, transported them by a form of anti-gravity, and assembled them quickly into complex and precise structures. Conventional sources assert that the Great Pyramid was built as a tomb for the Pharaoh Khufu, but this has been questioned recently. Only a few inexpertly executed hieroglyphic inscriptions were discovered in this pyramid, and these were believed by some experts to be fraudulent, dating from early investigations within the pyramid in the 1800's. The dates of construction of the pyramids are now thought by some experts to be well before the age of the later pharaohs. From its complex structure it seems possible that the largest pyramid was intended principally for some technological purpose, such as energy generation or signalling off-planet, and was not intended principally as a tomb.

Ancient Astronomy

Plato stated that the ancient Egyptians had been observing the stars for ten thousand years, but Graham Hancock remarked on the apparent anomaly that sky-gazing was a task for sea-goers, so why were the Egyptians, a landlocked people, so obsessed with astronomy? He noted that ancient seagoing vessels were found buried near the pyramids, suggesting that an earlier mariner race had introduced the Egyptians to this obsession in remote prehistory. They then took this obsession to levels of refinement only equalled by the much later study of this phenomenon by our own civilization from several hundred years ago. But this obsession came from where? From Atlantis? This is the only pre–Ice Age civilization of humans we seem ready to hypothesize as a progenitor of city life, and recent discoveries show that a technically advanced society *did* exist in Antarctica before their continent was moved to the South Pole and iced up. These recent discoveries suggest some interesting linkages with the Sumerian story and other ancient societies in the southern hemisphere, but we cannot now take for granted that the majority of the population of a more clement Antarctic continent was human.

The astronomical knowledge of the Sumerians, pharaohs, and Mayans could have come from training by a more civilized source with long experience and access to celestial observations. This requirement appears necessary in order to avoid assuming that these new human cultures could have invented a complex system of astrological time division, which logically must have been derived from observations in space over preceding millennia before Sumerian, Egyptian and Mayan civilizations came into existence.

The 365-day approximation to the year by the Mayans is an achievement one might expect from our short-lived species, but what is not so obvious is their long calendar,[66] which documents years back to 3,114 BC, preceding their civilization as we know it. In the case of the ancient Egyptians, the long celestial cycles in their original calendar shift from one zodiac house to another every 2,160 (= 5 x 3^3x 2^4) years, and form a cycle within the even more extreme 25,776-year 'Precession of the Equinoxes' or 'Great

Year'. These are features one might only expect to have been developed by a long-lived and ancient race and not by a species with a lifespan of not much more than three-quarters of a century which was just starting to experiment with city life. And why on earth would the Sumerians invent a time unit called a *shar*, equivalent to 3,600 years? This unit could have utility for the millennial Anunnaki culture on Earth, given that supposedly this is the orbital period of Nibiru. Its utility for describing day-to-day events of relevance to short-lived human societies seems improbable.

Searching for a diagnostic of our lifespan and how it affects our perception of history, it was pointed out in an Italian newspaper today that 700,000 people in the suburbs of Naples live on the slopes of Mount Vesuvius and that there is no main highway to help them rapidly escape a major eruption, such as occurred in AD 79, and destroyed Pompeii and Herculaneum and their populations. Vesuvius has erupted about three dozen times since AD 79 (i.e., roughly once every half-century since then).[77] The most recent eruptions were from 1913 to 1944, but there was an explosive eruption in 1631 that killed four thousand people. Despite optimism that the current eruptive cycle has finished, Vesuvius is an active volcano, and it almost certainly will erupt again.[77] We may conclude from this account, that historical facts soon become less important mythologies for a short-lived intelligent species. What we are seeing here—and elsewhere in human societies—is the rapid decline in importance of earlier historical dates with successive generations. This is often referred to by economists as 'future discounting'. What seems to emerge from this discussion is not the improbability of excessively long Anunnaki lifespans, but the remarkably short lifespan of the human species, and what this implies for our impact on the planet. For example, the last half century has seen the extinction of thousands of species due to the absence of conservation measures of much too short a duration, which in some cases were cancelled due to the contrasting beliefs of successive political regimes.

A generational interval for humans is 25-30 years, and, in practice, this is often the time interval used for discounting past events which are likely to have important effects if repeated in the future. This means that an event one hundred years ago that took place in the lifetime of your great-grandfather, (or great-great-grandfather, depending on how old you are), has lost much of its critical importance for you by now. Nonetheless, it certainly seems unlikely that an individual with a lifespan of one thousand years would choose to live near Vesuvius!

Evidence for Space flight Reported by the Sumerians

Sitchin has provided more than enough evidence to support his hypothesis of the Anunnaki arriving from off-planet long ago, and for a detailed description of his findings please see his publications. Nonetheless, I will report on several points that for me were crucial in agreeing to his main conclusion. First of all, four thousand years ago the Sumerians had identified all the planets around the sun, even the far distant ones, which were only rediscovered by us in the last century. Perhaps most convincingly, however, the number seven was assigned to planet Earth, which is only valid if counted in sequence from the furthermost planet moving inwards towards the sun, as you would expect if this information was imparted by an extraterrestrial tutor coming from the outer fringes of the solar system (figure 17). The fact is that in Sumer, and afterwards in Babylon, an annual festival was held resembling somewhat in its structure the Christian ceremony of the Stations of the Cross. This celebrated the first voyage of the Anunnaki from Nibiru to Earth, with scenarios for each planet they passed. These were apparently enacted based on the spiritual characteristics of the god which each celestial body was supposed to possess (e.g., the Sumerian equivalent of Gaia in our case?).

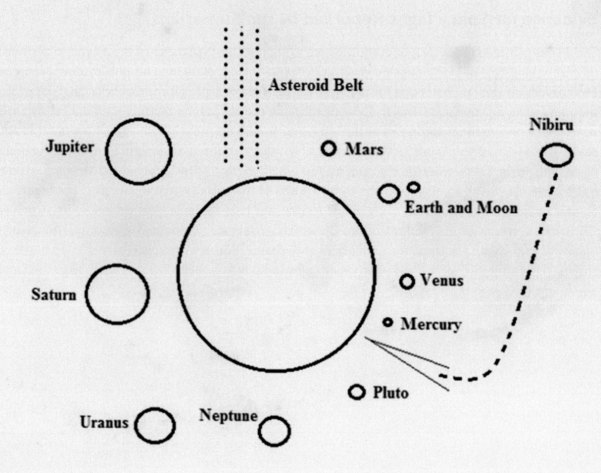

Figure 17. Representation of Sumerian knowledge of the planets of the solar system, showing their relative size but not their position. Those inside the asteroid belt, notably Mars, Earth, the Moon (which they considered a planet), Venus, Mercury, and the Sun (also counted in their planetary series), were considered separately from those outside the asteroid belt. In total, for each of these twelve bodies, a god was considered to exist.[1]

The 'righteous ones of the bright pointed objects' referred to by ancient Sumerians in figure 18 evidently described the crew of the Anunnaki space vehicles, and it is difficult to imagine alternative meanings for the symbols shown in figure 18 that could so closely resemble space vehicles in actual appearance! Thus,

Sitchin noted that several symbols from early Sumerian writing clearly identify space vehicles, also shown in figure 18.

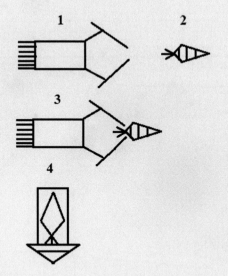

Items apurtaining to 'The righteous ones of the bright pointed objects' (after Sitchin):

1. DIN.GIR: 'The pure ones of the blazing rockets'

2. GIR : 'Sharp-edged object', or 'Rocket with compartments and fins'

3. Illustrating why 1. is open at the apex: it is presumed that 2 is intended to be an exploratory module attached to the main rocket.

4. KA.GIR: 'Rocket's mouth' (i.e., rocket in launching module)

Figure 18. Sumerian symbols referring to space vehicles (after Sitchin[1]).

Figure 19 shows that the launch module (also no. 4 in figure 18) was built adjacent to the temple for the god in question (i.e., in an enclosed sanctuary where humans were prevented from entering, being denied access to space vehicles). In fact, the construction of a launch platform by humans after the Great Flood, was mentioned by Sitchin as the motive for humans to construct what has become known as the Tower of Babel, for reasons that relate as much to its function as a lighted beacon to guide space vehicles while landing, as for its ceremonial significance. This proposed action, together with their intention to construct a space vehicle, was supposed to have led to the decision of the Anunnaki to disperse groups of humans to different areas of the world, where linguistic communication between them would have been difficult. Thus, if they sought to contest the power of the extraterrestrials, a strategy called 'Babel', which was obviously concocted by a very long-lived dominant species, would eventually have prevailed over an Anunnaki time frame, and would have led to a culturally and eventually linguistically separate races. Attempts by historical linguists to trace back the evolution of languages seems to demonstrate that language diversity was less evident in the past. Could this forced geographical diversification of human communities have provided the basis for subsequent interregional conflicts between human cities and states?

An early Egyptian wall painting[61] provides evidence for the use of rockets and space flight at the time of the ancient Egyptians—this would have been a result of Anunnaki connections with proto-civilizations on the river Nile in the pre-Pharaonic stages of this great civilization.

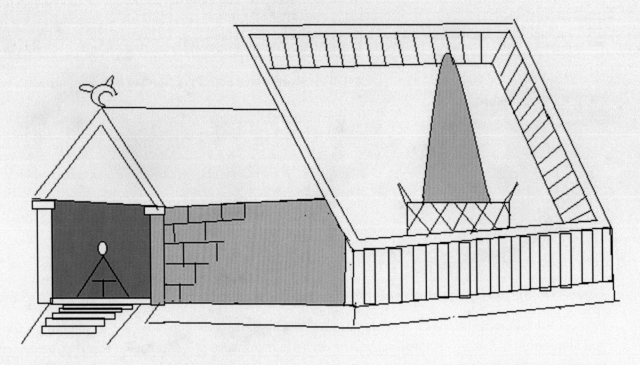

Figure 19. Drawing based on figure 67 in Sitchin's work[1] described as the installation of a *mu*, an oval-topped conical object kept out of view of the Sumerian populace in the inner sacred enclosure of the Great Temple of Ishtar. Despite controversy on this point, Sitchin indicates that this could be the 'boat' the goddess used to 'roam the skies, far and high'.

Space Flight as a Means of Escaping Catastrophes?

One objective of space flight is to access other planets, most likely those without a breathable atmosphere or no atmosphere at all. The easiest place to avoid problems on an airless planet and conserve a precious breathable atmosphere is below ground. The fact that some extraterrestrials (e.g., the Greys) have large eyes could be an evolutionary result of living below ground on a planet without atmosphere, i.e., in low illuminations when not involved in space voyaging. Perhaps underground is where long-lived, intelligent species take refuge from planetary catastrophes? Occasional reports of recent underground encounters with ET species have given credibility to this strategy of avoiding cosmic collisions or encounters with human populations.

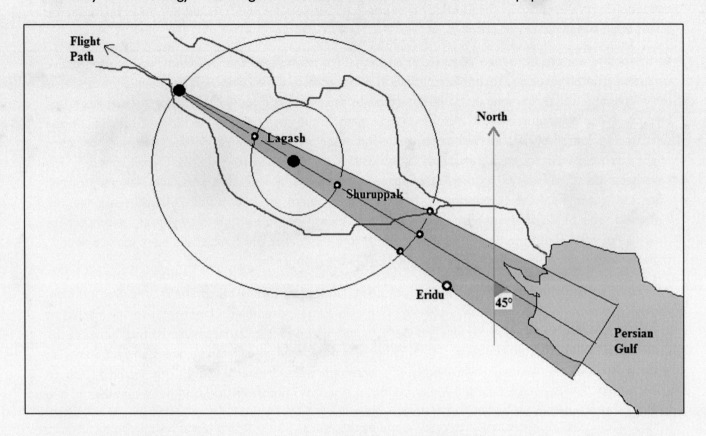

Figure 20. Approximation to figure 132 in Sitchin[1] he copied from a scratching on a ceramic surface found at Susa dating from about 3,200 BC. A reasonable conjecture by Sitchin is that this shows the flight path of spacecraft coming in to land in Mesopotamia. What was also implied by the original was that the positioning of Sumerian cities (circles), and presumably their ziggurats or other light sources used for spaceship guidance, were pre-planned for security in landing their space vehicles.

Of course if your lifespan is measured in thousands rather than hundreds of terrestrial years, you could have come to view life on the surface of a planet as a risky proposition. After all, a negligibly low annual probability of a catastrophe becomes close to a certainty over a geological period. In addition, our current dependence on the photosynthesis of plants exposed to the sun (for food) would not be a constraint for scientifically advanced organisms.

At this point I want to introduce a cosmic criterion that has been suggested for identifying the progressive evolution of intelligence and technological skills in extraterrestrial civilizations. It has been proposed that technically advanced civilizations should be detectable by astronomers from their works of 'Astroengineering'. Hypotheses on the occurrence of 'alien mega-structures' in the cosmos go as far back in science fiction as Olaf Stapledon in the 1930s. In 1960, Freeman Dyson described how technically super-advanced societies might surround their stars with enormous solar collectors. A star hosting such a 'Dyson sphere'[47] may have been identified recently by the Kepler space telescope, or at least that possibility has been raised. I find it doubtful from the description of their climatic problems on Nibiru that 200,000 years ago, the Anunnaki could be described as a category II species. Their manual approach to mining, and even perhaps their use of what appear to be chemical propellants in their rocket ships, may have been more primitive than the means of propulsion of contemporary UFOs using antigravity.

According to Graham Hancock, an ancient statue at Tihuanaco, locally called *The Friar* is dressed from the waist down in fish scales. It was believed that this statue resembled Oannes, leader of intelligent amphibious beings, and the Chaldean scribe Berosus recounted seeing Oannes while visiting Sumer in the distant past. This suggested connection with Sumer was reinforced when Sitchin quoted ancient Sumerian texts, saying that the Anannuki who navigated spaceships 'were dressed as fish'.[1] This would appear to confirm that *The Friar* was a bearded Anunnaki, or an Anunnaki-human hybrid, and not a member of one of the South American tribes who were notably beardless.

What Do We Know about Cosmic Archaeology?

So far as I know, like me, Knight and Butler were not specialists in the relevant sciences of astronomy, anthropology, or archaeology, but they based their conclusions on measurement units which are potentially repeatable, so it is hard to see how their evidence can be set aside. On the other hand, exploration of the solar system may soon provide us with new information on cosmic archaeology from ground-truthing on the Moon and other planets of structures where photographic records already exist. These photographs (see e.g., YouTube sites), show hard-to-deny evidence of artificial extraterrestrial structures on the Moon and Mars which are presumed to be much older than those from human civilizations on Earth. Ingo Swann[78] asserted that the Moon is much older than our planet and inhabited by space-going species. If so this may account for the sudden cessation of the Apollo missions to the Moon some forty to fifty years ago.

Artificial structures on moon?

Certainly structures and artefacts on airless planets are protected from the erosion and disintegration our active climate and oxidizing atmosphere inflict on ancient structures on Earth. The moon, in particular, is well-endowed with impressive structures, such as the Shard, a mile and a half high.[78] Hence, off-planet archaeology should provide us with vital clues as to our own distant past if we are allowed to go there, and as a person with a broad range of interests, I am impatient to learn more about this! I will also take the risk of mentioning here the ruins of miles-high fragile glass structures shown in photos taken by American astronauts on the moon, being confirmed as well by photos taken by the Chinese planetary vehicle Jade Rabbit deployed there in 2013.[59] These glass structures were presumably occupied many ages ago by an advanced extraterrestrial species. Watching our species during our evolution might have been one of their objectives.

It may be useful to speculate briefly on the climate of the planet Nibiru. Since it is postulated to spend several thousand years distant from the sun during each orbit, logically its atmosphere should be frozen in darkness as on other outer planets. Nibiru was described by Tellinger to be a large planet with a warm core and active volcanoes, giving off some form of radiation. How a solitary planet could provide a source of heat as well as radiation for vision and for photosynthesis on its own surface from volcanoes alone, is not obvious. When we examine the Sumerian symbol for Nibaru (Fig 21) we see a cross, which implies a planet crossing the orbit of other planets, showing wavy lines for radiation. It appears that this supports an alternative interpretation—namely, that we could be dealing with a dwarf star associated with Nibiru, the planet in close orbit around it. This is even implied by the Anunnaki symbol for their planet (figure 21), if we assume that with this symbol we are viewing Nibiru (the large circle in the distance), from behind the dwarf star (the small circles in the centre), which is assumed to be giving off intense radiation. Without a modern telescope, it would probably be difficult to distinguish this cosmic pair from a single radiating body.

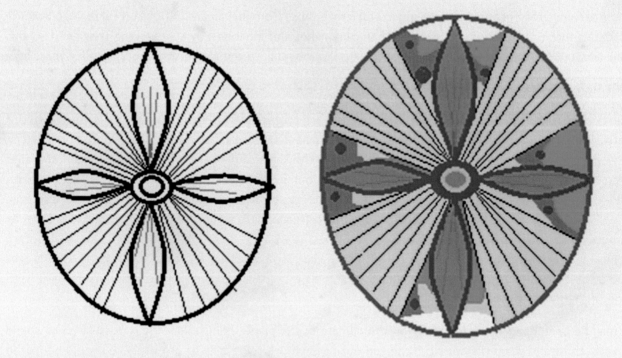

Figure 21. (Left) A pen and black ink version of the Nibiru symbol portrayed in Sitchin's book.[1] (Right) An imaginative colour version of the same figure, assuming the central circles are the active dwarf star emitting radiation (yellow rays), and a cross of red arrows of perhaps symbolic significance may refer to its crossing of the orbits of other planets while near the inner solar system. The planet in the background, behind the dwarf star in the coloured version, is a portrayal of Nibiru (as imagined by the author), with water bodies (blue), dry land (brown), and green for hypothetical vegetation, with volcanoes (red) and ice caps (white).

Some of Sitchin's Sumerian clay tablets suggested that it had become urgent for the Anunnaki to find a solution to the heat loss occurring from their home planet, and as mentioned earlier, they were aiming to counteract this by dispersing monoatomic gold particles in their upper atmosphere. A dwarf star in proximity to the planetary surface was presumably the source of light for agricultural activities during the millennia spent far from the sun's radiation, and this heavy dwarf star would be the gravity source which on occasions was postulated to lead to disruptions on other planets of the solar system during its sunwards migration.

Elsewhere in the solar system, there is evidence that intelligent creatures were busy in the distant past, and may still be there. However no mention of this emerges from terrestrial authorities who may wish to keep information on other intelligent creatures secret, at least until military action is decided upon. Incidentally, of relevance here is a registration made on the moon by equipment left on the surface by the astronauts to record geological vibrations or movements of the moon's surface strata. When the *Voyager* launching structure was allowed to fall back onto the surface of the moon, it registered a long, sustained resonance lasting several minutes. This led to speculation that the Moon is hollow and, from other evidence, inhabited. On Mars, ribbed 'transportation tubes' still protrude above the surface,[56,57] and there are other mega-structures there that resemble ancient Egyptian edifices, as well as large 'portraits' of individuals engraved in the planet surface, some of whom resemble humans. Like the Nazca images, these can only be appreciated from space, and the construction of such 'markers' to help spacecraft locate isolated settlements is of obvious utility for a space-going species. No other use is evident, given that the images are not visible from the planet's surface. This interesting functional feature of ancient landscape art is evident on both Earth and Mars, notably, large-scale images of faces, geometries, and other symbols cut into the planetary surfaces, As has been suggested for the Nazca imagery, the 'site identifiers' for incoming spaceships approaching a colony on an unfamiliar planet, provide the pilot with a recognizable landing location visible from orbit. On a much smaller scale and more recently, crop circles which demonstrate geometric concepts have been impressed on corn fields in England and elsewhere, presumably for our education.

One wonders if a linkage was established with Earth by a former Martian civilization? This may seem unlikely, given that their civilization long preceded humanity. Nevertheless, symbols such as 'The Face' at Cydonia, Mars, have been a subject of controversy between the growing number of those who accept extraterrestrial remains exist on Mars and those who believe that this image is a trick of light and that we are the only intelligent species to evolve in the universe. Or are we looking at the results of the activities of those Anunnaki personnel referred to as the 'Watchers', who were stationed there to aid in transhipping materials between vehicles from Earth and long distance carriers to Nibiru? The suggestion from Sitchin's work is that the 'Watchers' on Mars transferred gold shipments to larger freighters travelling to Nibiru and had built a local community on Mars for this purpose that supposedly lasted at least a thousand years according to some opinions. The giant images constructed on the planetary surface there might have been intended to communicate with their fellows on Earth? If so, had they also experimented with the building of pyramids on Mars before finalizing a design for their construction along the River Nile? The design of the Gaza pyramid shows evidence of technological sophistication clearly beyond the technicalcapabilities of early human civilizations, however, such a hypothesis of an off-planet linkage would require moving the dates of pyramidal structures on Earth back to a far earlier era from the construction dates now given of around 2,000–2,500 BC, to accommodate such an apparent coincidence. Sumerian inscriptions also suggest that the early Egyptian civilization could be linked to an exodus from the South African mining complex some 50,000–100,000 years previously, but the original Martian civilization seems to have considerably preceded this date.

Recent doubts have also been expressed as to the dating of the monument called the Sphinx constructed adjacent to the Nile pyramids. It should be borne in mind that vertical runnels down its sides are suggestive of ancient water erosion. From the climatic history of Egypt, this could not have happened later than 7,000 BC, when regular rainfall effectively ceased in Egypt (i.e., its construction may have occurred some 7,000–9,000 years ago). Did its builders leave us any evidence of its origins, hidden somewhere in the vicinity, as some have suggested?

Ancient Nuclear Incidents on Mars and on the Earth

The detection of an artificial nuclear isotope Xenon 129 in the atmosphere of Mars, and a significant concentration of uranium and thorium in its surface soil were reported in a 2014 paper entitled 'Evidence of a Massive Thermonuclear Explosion on Mars in the Past' by a scientist, Dr John Brandenburg.[68] He noted that the isotopes present resemble those left after hydrogen bomb tests on Earth and suggested that these conditions were likely the result of two air blasts of large nuclear ordinances, traces of which are apparently still evident on the northern hemisphere of the planet. In a video, he estimated these explosions to have occurred between 500–200 million years ago in an area where evidence of ancient constructions was apparent. Mars may therefore present an example of an early civilization wiped out by a nuclear attack from space, losing also its water bodies, atmosphere and life forms.[53,54] By whom? And did any survivors arrive on our planet afterwards?

The resemblance of some of the structures seen on Mars to the pyramids of Egypt, as noted, is paradoxical (Fig 25). A native Martian civilization was evidently much earlier than either the birth of humanity, or the later presence of the Anunnaki 'Watchers' on Mars. The occurrence of large-scale engraved 'faces' on Mars and smaller 'statues' picked up on photos by the US *Rover* urgently require further study. The suggestion already made to explain the large-scale engravings in the Peruvian desert near Nazca may also be relevant elsewhere for identifying planetary communities from orbit; by whom, and when?

That the Anunnaki brought with them advanced weaponry from Nibiru which was first hidden, but later used, was demonstrated in the translations of Sitchin. Another, perhaps related, and technically advanced early human civilization on the Indian sub-continent, also had warlike traits and futuristic weapons as suggested by ancient Sanskrit texts quoted earlier which described aerial warfare between their 'gods'. Fused circular platforms of glass were left on the ground at sites where air blasts of recent atomic weapons were tested. Similar platforms of fused green glass have been found on the ground in India, Africa and the Middle East,[69] and it is not unreasonable to assume that they may mark points of atomic explosions in the distant past, some of which are also confirmed by the evidence presented in ancient Sumerian[1] and Indian texts. In *Veda Knowledge Online*, when he was asked whether the Alamogordo nuclear test was the first atomic bomb to be detonated, J. Robert Oppenheimer, the scientific coordinator of atom bomb development before the Second World War, replied: 'Ancient cities whose brick and stone walls have literally been vitrified, that is, fused together, can be found in India, Ireland, Scotland, France, Turkey and other places. There is no logical explanation for the vitrification of stone forts and cities, except from an atomic blast'.

A comprehensive review by Steiger suggested that the epoch previous to the last Ice Age was one where many cities in India and elsewhere were destroyed together with their inhabitants. It is to be fervently hoped by all rational individuals that a repetition of these horrendous events on the subcontinent and elsewhere can be avoided in this epoch! Although Oppenheimer's comments may explain Sumerian accounts that atomic warfare destroyed some early cities, notably Sodom and Gomorrah in Mesopotamia, vitrification of massive stone blocks in pre-Inca ruins may not have been an effect of a nuclear explosion, but a feature of an unknown megalithic construction technique whereby the rock is softened by heating or by a 'rock softening liquid' as speculated by some authors, to improve the fit to adjacent units. Some have even speculated that the use of metal pins between mega-blocks was to hold them in place while semi-liquid.

A related mysterious phenomenon Oppenheimer mentioned is the presence of vitrified forts on the west coast of Scotland and cliff-top forts on the west coast of Ireland. This could be evidence of very organized early civilizations in Great Britain and Ireland, dating from before 1,000 BC [50]. If so, there must also have been a maritime or space-based civilization present at that time that was implementing naval attacks against them. (Note that the inundation of ancient coastal cities by a 400-plus-foot sea-level rise during the Great Flood, rather than nuclear explosions, was probably the main cause for their disappearance). Nonetheless, one may wonder if the ready use of nuclear weapons in war as a solution for political problems by the Anunnaki and their hybrid descendants, has had an influence on our own approach to similar difficulties?

The 'Handbag of the Gods' Seen in Visits to Several Primordial Cultures

Graham Hancock travelled worldwide from Mesopotamia to the Americas and found that the Sumerians, the Olmecs, the Mayans, the Aztecs, and other ancient civilizations displayed the same sculptural or engraved stone motifs when depicting honoured visitors, who were often later adopted as divine beings. Two common objects portrayed in sculptures in these different locations were the mysterious 'bag' they all carried (Fig 22), and the 'wrist watch' seen on nearly all depictions.[43] That these were a focus of attention implied that those visited were told their true function which must have impressed them.

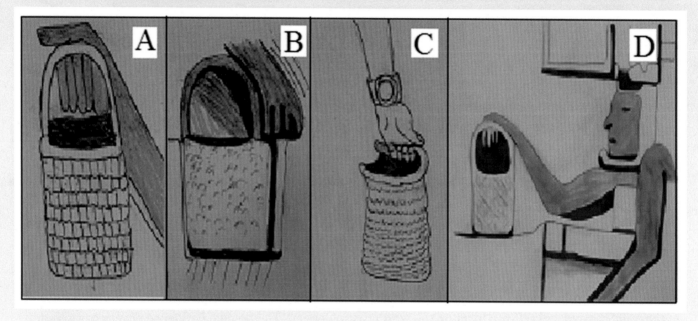

Figure 22. A key point made by Ivan, editor-in-chief at ancient-code.com, and illustrated here by 4 drawings, is that the Anunnaki of ancient Sumer (C) visited (A) the Olmecs, (B) the 10,000-year-old temple of Göbekli Tepe, in Turkey, and (D) the Mayan civilizations. As shown in engravings of the 'visitors' or 'emissaries' at each location, all of them were equipped with the same mysterious handbag.

Apart from unresolved questions on the purpose of these important 'artefacts', the fact that they were carried by their visitors was considered notable by those visited. It also identifies them as all coming from the same culture—presumably from Sumer! This has implications for what we can deduce about

Sumerian/Annunaki travel throughout the world—by ship or perhaps more probably by flying vehicle? And could these visits be seen as contemporaneous and altruistic as suggested by Hancock, in seeking to pass on knowledge of technologies lost by those visited by the Great Flood? Or perhaps they were prospecting for gold which is found in all the locations mentioned. What was in the bag requiring its owner to clutch it so closely? Remote controls for his vehicle? A weapon? Gold-prospecting equipment? The last image in figure 22 may suggest one application, by showing a seemingly Mayan figure crouched over the controls of what appears to be a space vehicle with an eagle's head above it (shown in the original). He has a technologically complex helmet on his head and is clutching his handbag in his right hand as he peers ahead of the vehicle. Taking into account the recent development of mental guidance systems in modern jet fighters, could this suggest that the bag contains equipment linking him to the sensors of his vehicle or its mental controls?

Some Ideas for a Terrestrial Sequence of Events

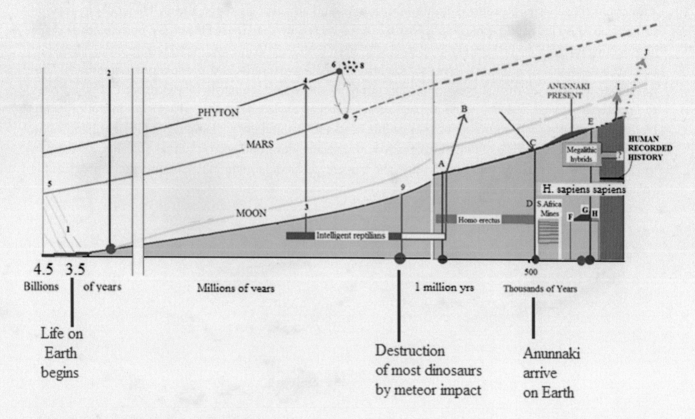

Figure 23. A personal hypothesis of events that may have preceded human evolution to intelligence according to various sources. Few early dates are available and those shown, (and often the events themselves) are very, very provisional! The right-hand (brown) section with a finer time scale shows the duration of recorded history, while earlier events and entities that may have impacted the ecosystem of planet Earth over longer time scales are shown in the two sections to the left with progressively longer (and very uncertain) time scales. (See numbered key below for details). Red circles suggest genetic modifications to pre-humans. Purple circles suggest planetary disasters. Arrows signify hypothesized space flights.

Key to figure 22. (Note: Numbers referred to on this figure are not those used for references).

(1) Fossilized micro-organisms have been located in meteoritic material before and after they landed on Earth, and more recently, metal micro-capsules have been located at 40 km altitude in the stratosphere containing bacteria, supporting the hypothesis of directed panspermia. This would have speeded up evolution and might even have encouraged convergence of morphologies on different planets within the solar system. (2) Collision of the proto-Earth with another planet may have created the moon or captured it as a satellite from the other planet. (The possibility of its artificiality has also been suggested). (3) Perhaps there were intelligent reptilians who left the Earth before the mega-asteroid impact 65 million years ago? (See point 9 below). Did intelligent reptilians emigrate to planet Phyton? (5) Evolution to intelligence began much earlier on Mars than on the Earth: are the survivors still there underground?

(A) An expedition from the Pleiades arrives in Earth orbit. According to Fenton, their interstellar vessel was destroyed, but the survivors worked on proto-humans to further their mental/physical development. The reptilians are banished from Earth. (B) The Anunnaki arrive on Earth and remain here for several millennia, especially in Mesopotamia and southern Africa, but they also visit the Americas and the pre-Antarctic continent where they established communities prior to its movement to a polar position which destroyed their base on Earth. A transmission centre on Mars was established for reloading gold onto long-distance freighters for Nibiru. Just before, and immediately after the last Ice Age, they (or their human hybrids) were probably responsible for megalithic architecture worldwide, using Anunnaki technologies we no longer possess. (C) The human workforce is upgraded by the Anunnaki using gene splicing on earlier humanoids. (D) After modification to improve their capabilities as a workforce, humans are employed as labour in gold mines in southern Africa from around 200,000 years ago. (E) Human hybridization with Anunaki results in 'cone-head' giants and the megalithic civilizations. (F) During the melting phase of the last Ice Age, many civilizations established below current sea levels (then, 400 ft lower) were inundated and disappeared. (G) A sudden movement of the Earth's surface crust resulted in a rapid displacement of continents relative to the North and South Poles: The Antarctic continent shifts to the new South Pole. Recent evidence suggests that, previously, a civilization of the Anunnaki existed on the Antarctic continent under more temperate conditions. (H) The Great Flood inundated antique coastal cities in many civilizations (e.g., India, Japan ... Atlantis?). (I) Humans travel to the moon. (J) Humans now propose to voyage to Mars.

The diversity of potentially intelligent hominid species coexisting on the Earth in past ages has been illustrated by occasional archaeological and anthropological discoveries. This diversity could have been achieved slowly by natural selection alone, but it could also have been speeded up if a more advanced species, then visiting or resident on Earth, had chosen to 'improve' our genotype. Obvious strategies such as gene splicing and clonation of chosen physical types of hominids would certainly have been within the technical capacity of ETs. If this was the mechanism used, it likely first incorporated a wide geographic selection of genetic strains, perhaps then bringing different groups together for cross-breeding, aiming for maximum intelligence and domesticability. This last characteristic was also speculated upon by Colin Barras,[78] who asked, "Most domestic species were tamed by humans. So what tamed us?" He quoted work speculating that the relevant genes associated with 'tame' behaviour in domesticated animals are linked to the neural crest and suggested that early in our evolution, 'humans underwent the same sort of domestication as these animals did' but considered it to be self-domestication, thus allowing humans to cooperate non-aggressively in larger communities than small hunting bands.

The Anunnaki may have had the same objective in mind, namely to help tame us by 'improving' the human breeding stock so that we would be able to accept orders. (And perhaps to extract genetic material for their own use?). One could, of course, vigorously deny that our species was used in this way, but we ourselves have used this mechanism in breeding 'inferior' species, just as we may have been referred to at the time by super-intelligent extraterrestrials. The final result from a widely branched hominid evolutionary tree is strange but also compatible with this hypothesis, namely, that only one species survived a long period of diversified evolution! On this point I think it would be unusual for land mammals with a global distribution for only one species to survive after a million years of evolution in widely different environments. What was the dominant mechanism: 'ethnic cleansing' and 'forcible crossbreeding by an aggressive variant'; a friendly exchange of genetic material between different tribes; or was it a systematically guided form of rapid evolution by extraterrestrial species, using crossbreeding, cloning, or gene-splicing?

What Was the Relative Timing of Hominid Affairs in the Solar System?

There is a wholly inadequate sum of data available to draw any firm conclusions on the timelines or even sequences of key events early in hominid evolution and associated events in the solar system. It is intriguing, nonetheless, to consider the evolutionary tree of terrestrial hominids, as far as we believe we know it, alongside other events which may have been going on at roughly the same time. Figure 16 also shows the limited cross-breeding that is believed to have occurred between humans, Neanderthals, and Denisovians, and this may not have been necessarily distant in time from the period of the Anunnaki genetic interventions on *Homo erectus*. On an even earlier time frame, our distant ancestor, *Homo habilis*, may perhaps have been aware of wars in heaven that after thousands of generations left their offspring with a fair share of the future solar system?

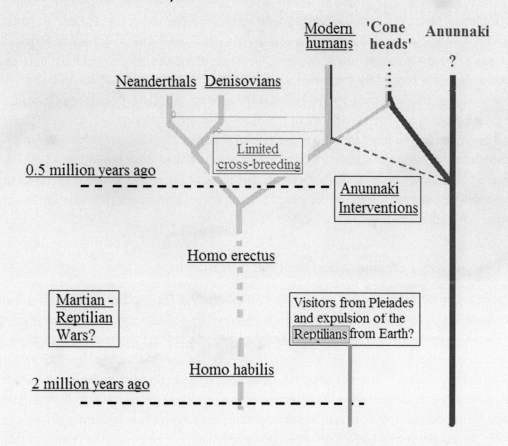

Figure 24. Placing on the same uncertain timeline the evolution of our species and our hypothesized interactions with other intelligent species mentioned in the text. Issues hypothesized here are: (1) Limited cross breeding of 3 terrestrial hominid species (2) the Anunnaki genetic intervention on Homo erectus, and 3) hybridization producing the 'cone heads' (4) much earlier still, the hypothesized expulsion of the then dominant reptilians from Earth, and (5) the war between the original Martians and the reptilian inhabitants of the

planet which was pulverized to form the asteroid belt. The occurrence of regional genetic diversity in modern humans does not result in our mutual infertility, and hence must be relatively recent.

Evidence from human history is confined mostly to events since the last Ice Age, hence opening up our assumptions a little wider in this debate may be useful. The usual assumption is that the further back you go, the more primitive the peoples you encounter. Unfortunately, the sources of hard data to support this conclusion before the last 4,000 years or so are few. This makes it difficult to automatically debunk theories (like this one) that are based partly on DNA records, partly on evidence from excavations and sculptures and from the written records of the Sumerians and other ancient civilizations. From the archaeological perspective, the few ruins and fragments of traditions and artefacts from before the last Ice Age are dominated by mysterious constructions of closely interlinked giant stone blocks, often weighing hundreds or even thousands of tons. The mysteries posed by these very ancient structures tend to be ignored by specialists, perhaps fearful of the impact that information-limited speculation on prehistory would have on their professional reputations.

A Personal Hypothesis on Human Historical Landmarks

A significant proportion of this document was written on the basis of reports from the internet, and as such, will not be considered an accurate piece of historical research by the experts! It does however, provide a possible framework, some questions and options for future hypotheses and some fertile directions for future studies. At this point we need to try and form a 'bridge' between the isolated ancient facts that remain and the creation of a complete picture. I have not cited references for all assertions made, although very few of the controversial points were my original ideas. As I mentioned earlier, I haven't tracked back all units of information to the specialist literature, but at the same time, no assertions have been made without a relevant item having been read in modern texts on the subject or on websites which appear to be serious in intent. Taking all this preamble as read, the following is my personal 'hypothesis' as to the events in the last few million years or so, and anybody is welcome to make the changes to this they feel are necessary.

In summary, I have adopted Sitchin's hypothesis that an extraterrestrial species, referred to by the Sumerians as the Anunnaki, landed in Mesopotamia some 400,000 years ago and cultivated a civilization with world-spanning connections in which the search for gold was a stated priority perhaps explaining our subsequent obsession with a metal which then had few practical uses. The Anunnaki were described as hominids resembling mankind, but up to 12 feet tall, with an elongated cranium. They came from a totalitarian monarchy with fixed social classes and a paternalistic attitude to their human subjects. Eventually this led to rule by human-Anunnaki hybrids, and finally, much later, by the humans they modified from early hominids by genetic manipulation and trained to be useful workers. Initially, their search for gold was concentrated in the Indian Ocean and the coastal seas of the Arabian Peninsula, but from roughly 250,000 years ago, there was a refocus on the southern areas of Africa. An extensive series of residences, mines, and associated agricultural terraces were created on the southern African veldt and adjacent hills. The purpose of this 'treasure seeking' as suggested by Sumerian inscriptions, was to locate this precious metal and slow down the deteriorating climate of Nibiru by adding monatomic gold to the upper atmosphere, given that their planet was becoming uninhabitable due to heat loss while distant from the sun.

Limited evidence supports the hypothesis that an extensive Sumerian/Anunnaki connection at other world sites existed where occasional Sumerian artefacts have been found. Notable among these is the 'Fuente Magna Bowl' found near Lake Titicaca, Bolivia.[70] The inscriptions within it have been identified as proto-Sumerian, referring to the Goddess Neith, a divinity very popular among the ancient Libyans before they left the region to settle in Mesopotamia and elsewhere. This find suggests a linkage between this area of South America, rich in precious metals, and the Anunnaki story. A linkage also seems to have been established recently to an unspecified kingdom on the Antarctic continent,[34] not then completely covered with ice, which may have been Anunnaki dominated. Its existence could have provided the basis for the Atlantis mythology, but perhaps not as a city of humans as usually imagined. Recent Internet reports suggest that the discovery of an abandoned city under several kilometres of ice, and several large pyramids protruding through the ice seen on Google Earth, illustrate that an advanced civilization once existed there. However, other extreme claims on the Internet as to newly discovered Antarctic phenomena implying an Anannuki prescence, is a story left to the readers to follow up on, once more detailed information emerges.

Highly relevant however in corroborating Sitchin's conclusions, is the recent discovery in southern Africa of the remains of an ancient city extending over 1,500 square kilometres, dating back to between 160,000 and 200,000 years ago judged from the rate of erosion of dolomite stone.[10] This seems part of an even larger area of archaeological remains extending over some 10,000 square kilometres. The organized nature of this ancient community and a road network connecting it to adjacent terraced agriculture, suggest that the metropolis was built by an advanced civilization, which was not believed to exist at that date on Earth. The majority of the inhabitants may have been upgraded hominids, kept in closed structures when not engaged in slave labour. From the presence of ancient mine shafts and given the local geology, a major activity of the 'residents' in gold mining was suggested. This seems to identify the city as an Anunnaki/

Sumerian colony with human laborers, and verifies information independently given on Sumerian tablets. Rock carvings of the Egyptian Ankh and other symbols, according to Tellinger [11] seem to anticipate by thousands of years an Egyptian connection for what he suggested is the oldest African city on the planet.

It emerges from the tablet accounts that the Sumerians held the Anunnaki in great respect. According to Tellinger, humanity was later guided into considering their guardians as deities, with all the subsequent consequences of obscuring perhaps, our understanding of historical events in areas that later came under the effective control of world religions. This explains In part the automatic assumption of early translators that the Anunnaki were 'sky deities', and it perhaps reflects the fact that in the late 1800s before the Wright brothers invented aeroplanes, any description of a physical entity descending from the sky was a fantasy. This was not the case for Sitchin, given the contemporary reality of human flight and initial space programs at his time of writing. There is adequate evidence on Sumerian tablets to support an interpretation of the writings of Sumerian scribes, by assuming the technological competence of the Anunnaki in space flight and in other medical, agricultural and social fields of endeavour, when you consider that their activities were later recorded as history by scribes with a limited familiarity with science.

The key question anyone must ask, is whether the rapid transformation of primitive hunter-gatherers into city dwellers supported by intensive agriculture could be achieved in a short period without external assistance? Adding the South African archaeological evidence to our early history and postulating a subsequent migration northwards to form the pre-Pharaonic civilization in Egypt, to me provides a convincing temporal coincidence between explained and unexplained stages in human evolution.

The Sumerian King List (see Wikipedia) is an important clay tablet which provides a chronicle of the kings who ruled in Sumer before and after the Great Flood. The list starts with this statement: 'After the kingship descended from heaven, the kingship was in Eridug. In Eridug, Alulim became king; he ruled for 28,800 years.' Even some post-Flood lifespans are given as lasting more than one thousand years. Difficult to believe, and there is the temptation to accuse the authors of gross exaggeration. In the Old Testament of the Bible however, we see a more modest increase in human lifespans reported, followed by a progressive decline in post-Flood human lifespans, from Adam's 930-years of life and Noah's 950 years to Abraham's 175.[24] Accepting these human lifespans as valid would require us to propose a mechanism for them. The fact, on its own, that they lived at a time before chemical radiation contamination of this planet, seems an unlikely explanation.

Differentiating between pre-flood and post-flood reigns and environments, a steady decline in the lifespans of Sumerian monarchs seems evident,[24] even for the Anunnaki. A likely physiological or genetic indicator of the potential lifespan of humans is the length of the telomere.[71] From a recent study of Steenstrup and co-authors in 2017, it does seem that this prolongation of the chromosome is progressively used up in successive cell divisions; hence, it forms a limit to our natural lifespan. Could genetic modifications by the Anunnaki have affected this microstructure so as to limit human lifespans to less than 120 years, and did this occur once the long-lived Biblical personages showed that the Anunnaki genes for longevity had been transmitted to a few early humans? If some of these early biblical figures were the descendants of humans containing Anunnaki longevity genes, this might explain human lifespans significantly longer than normal, even if historical exaggeration may still have played a part.

The lifespans of Anunnaki were documented by the Sumerians as extremely long, and this may have been a characteristic of the inhabitants of a cold planet with a dense atmosphere, and where a single year extends for 3,600 of our years, so that the inhabitants spend centuries outside the range of the sun's radiation. As a result there might be few short-term daily changes or seasonalities which could rapidly use up the physiological capacity for cellular regeneration and an advanced knowledge of the mechanics of cellular regeneration may have existed: but perhaps for the exclusive use of the rulers? What seems to emerge from the Sumerian accounts is the rigidity of inheritance of key positions in Anunnaki society. There seems to have been a dynastic form of monarchy, where the king was deified and supported by what resembles a caste system. Surely, extreme longevity as a natural phenomenon in such a society could be maintained only by a rigid control on the right to reproduce. Could this explain the eagerness of the Anunnaki workmen based on Mars to come down to Earth to reproduce with genetically-modified human females?

One interesting speculation reported by Tellinger, relates to the high percentage of 'junk DNA' in human chromosomes. As opposed to other species (e.g., 17 per cent for fruit flies) in our case for roughly 97 per cent of our DNA no genetic function can be assigned. It would be interesting to know however, what proportion of junk DNA exists in gorillas and other anthropoid apes, which I presume were not subject to Anunnaki attentions! At the same time, DNA analysis around the world is identifying unusual segments in the genome of local peoples that were not documented in the human genome analysis organized in the early 2000s. Whether it was *Homo erectus* who were the experimental subjects used by Anunnaki scientists close to half a million years ago, and not a range of hominid species is not clear, but Tellinger speculated that DNA from an Anunnaki source was spliced wholesale into the chromosomes of *Homo erectus*, and then the resulting genome was 'tampered with to allow only a small proportion of it to function'; i.e., under this hypothesis, a large proportion of Anunnaki genes were transplanted, but those for advanced functions such as longevity were subsequently eliminated, (possibly to avoid future competition with the 'master species' by their servant race?). Or were only a selection of Anunnaki genes transplanted? - and what did they leave untouched of our original genetic makeup as hunters and gatherers?

One convincing detail from the relevant Sumerian tablets describing this operation is that the first few attempts at gene modification were unsuccessful or produced defective offspring. I should add that it appears as if a significant proportion of native hominid tribes were unaffected by these experiments, but as illustrated by the deduction described for the Beaker People, subsequent exterminations of peoples who remained in the hunter-gatherer mode may have taken place. My personal impression is that what we refer to now as left hemisphere-dominated personalities were promoted by the Anunnaki genetic modifications; a personality type more adapted to implementing orders received from above without question.

Returning now to the Sumerian precursor record to the story of Noah's Ark. Enlil, the chief of the Anunnaki on Earth had reportedly become disgusted at the active reproductive behaviour of his modified human subjects, and perhaps this is why he allowed them to be swept away by the Flood? When he returned to Earth afterwards with his fellow species and found Utnapishtim (the equivalent of Noah in the Sumerian myth) and his family still surviving, he was furious with Enki for helping them build the Ark. However, his Nephilim advisors convinced him of the continued utility of modified humans as a workforce, given the need to grow food for themselves after the Flood had wasted the agricultural land they had developed. Hence, his memorable instruction to Utnapishtim, (perhaps with a touch of sarcasm?) was to 'go forth and multiply'.

Moving now to the period of the megalithic civilizations, a recent study of the genotypes of western Europeans has suggested that a massacre of the original male human populations in Spain, the British Isles and other areas of Western Europe, was responsible for the strange configuration of the modern Y-chromosome. This genetic analysis[86] postulated that the present configuration arose when western Europe was overrun by the invading armies of an extensive population from the east referred to as the Yamnaya. These people originated north of the Caspian and Black Seas, and became associated with the Bell Beaker Culture in their western migration. Their invasion of the Iberian Peninsula coincided with the disappearance of the local male line; only the males of the invading Yamnaya pastoralists left descendants. The radical interpretation of this invasion from the genetic data is that a violent conquest occurred, with total 'cultural cleansing', or more accurately, extermination of the local male population originally present.

It is presumed that the local female population survived to be taken over by the invading males. A similar violent conquest on British soils was also deduced by these authors [86] to have led to the close to extinction of the peoples responsible for building Stonehenge [103]. These traumatic events occurred some 4,000 years ago, and it may be reasonable to ask what aspect of their physical and intellectual capacity gave the invaders the advantage they apparently had over local populations? Could it have perhaps been associated with their upgrading from early humans by the Anunnaki, and that they were then 'programmed' and used for military action against other megalithic communities? This last hypothesis is a 'long shot' based on the discovery of cone-headed burials in the areas where the Yamnaya apparently originated. Since the Yamnaya also invaded much of Europe including the British Isles with the same strategy, they effectively became the ancestors of most modern Europeans.

From what we can deduce from recent studies of Stonehenge [103], its surrounding area was a large cultural center for much of the original population of southern England and adjacent territories, and was used for communal feasting and astronomical observations. That the males in this population may have been wiped out by what appears to have been an Anunnaki-trained and equipped invading force, suggests that this may have been the great tragedy that initiated future aggressions by modern humans in adjacent countries.

Given that the Anunnaki-trained humans eventually came into ascendancy, at least in Europe and presumably in unspecified parts of Asia and the Middle East, this might perhaps explain why warfare and colonial conquests fill our later historical record? In contrast, a natural human desire for a peaceful communal life, as illustrated by recent Neolithic excavations at Stonehenge, could have led to the birth of the type of religions which emphasize the need for higher sentiments in mankind.

One mystery remains, notably the occurrence of an Rh-negative blood group in 27% of the Basque population of northern Spain, who were believed to be among the first people to enter Western Europe, speaking an ancient language unrelated to other tongues. They may have descended from the Cro-Magnon stock. A view expressed in the first book of Basque grammar ever published asserted that Euskera was one of seventy-five languages to have developed out of the original single language spoken before linguistic diversification led to confusion at the Tower of Babel. [(92)]. Since individuals having the Rh-negative blood group have difficulties reproducing with partners of some other blood groups, the Rh negative blood group common in the Basques has been speculated to be of a different origin from other blood groups. If the RH negative factor does not derive from any known earthly link (being seemingly outside the theorized evolutionary process) – from where did it originate? Geneticists generally claim the RH-negative factor is a mutation of unknown origin which apparently happened only a few thousand years ago. Is this further evidence of early extraterrestrial intervention on our species?

John F. Caddy Ph.D.

The conclusion of this account of the Yamnaya invasion shocked me, as it will other readers. Going back to an earlier section of the text you will note that I personally share the dominant ancestry of most European men; thus my patrilineal ancestors were presumably responsible for the horrific extermination of men of the local tribal populations who were not 'upgraded' by the Anunnaki. On the matrilineal side of my personal inheritance however, after moving from Asia to Southern Europe in the Ice Age, the males on the matrilineal line eliminated by the Yamnaya were judged to be on the same line as the Solutrian Culture responsible for Cro-Magnon cave art. Although it is reaching too far to claim a personal connection in this case, my enthusiasm for the visual arts and poetry is shown on the 'Other Matters' section of my Google Cloud site*

(This of course is a feeble attempt to find an ancient 'family connection' for my artistic efforts, and to show my support for the hunter-gatherers!)

Nonetheless, my patrilineal ancestors were presumably descendants of the humans 'upgraded' by the Anunnaki. Athough we cannot say what character traits they acquired by this operation, one could speculate that these included an unthinking willingness to obey orders from authorities above them, even for so-called inhuman actions: but that they also showed elements of logical thought – i.e., features of left brain dominance.

The Mars Nuclear Incident

Two technologically advanced protagonists in the distant past, the original Martians and the reptilians, were visited in a remote viewing exercise before nuclear war between them apparently destroyed both their planets or rendered them uninhabitable. The distant viewers were also aware of a representative of a higher cosmic civilization present, who was attempting to defuse the conflict. Assuming that the highly evolved species this entity represented is still concerned to replace two intelligent species lost from the solar system, this could explain why preparing the Earth for the evolution of other intelligent species should have acquired some relevance for them, as implied by the work of Knight and Butler. The scenarios of nuclear radiation, high temperatures, pollution, and occasional geological catastrophes characteristic of our planet, could explain why the Anunnaki did not persist in migrating here, and why they did not seek to suppress human expansion. It must have been clear to them that without recourse to outright extermination which was against their principles, their species could not compete with a short-lived, highly fecund species subject to population explosions.

(*FOOTNOTE (https://sites.google.com/view/john-caddy-fisheries-science/)

These events took our species from perhaps a few million in 2,000 BC to 6.5 billion some four thousand years later. We now face the urgent problem of how to control our own populations in a world facing shortage of food and space at a time of chemical contamination and rising CO_2 levels and temperatures. From this perspective, the exercises in remote viewing mentioned in this account may provide us with some useful precautionary lessons.

Remote viewing, a technique developed by the CIA for military purposes in the Cold War, has begun to be accepted more widely as a means of investigating events distant in time and space. Although its underlying mechanism has not been established, it appears to confirm that the universe stores reliable information, and that details are available to the trained mind on events and structures far away in time or space. As mentioned, one remote viewing exercise reported [60] threw some light on the site of a disaster long ago represented by the asteroid belt located between Mars and Jupiter. It may also throw light on the disaster just recounted on Mars. The original intention of this remote viewing exercise was to explore the origins of the asteroid belt far back in time.[60] Its results suggested that the asteroid belt consists of fragments of an earlier planet Phaeton, previously inhabited by intelligent reptilians, which lay between the orbits of Mars and Jupiter. Its destruction, as observed by remote viewing, is supposed to have occurred as a result of nuclear warfare.

As mentioned earlier, the nuclear blasts on Mars were identified from isotopic evidence, and as a factual event this seems to add validity to the remote viewing study, suggesting that a war did take place between the Martian population and the reptilian inhabitants of Phaeton. The destruction of Phaeton was supposedly in retaliation for the bombardment of Mars by atomic weapons, an event which apparently did occur. This remote viewing session suggested that the reptilians destroyed the once viable aquatic and terrestrial ecosystems and atmosphere of Mars, perhaps because the original Martians, another scientifically advanced society, refused them the rights to mineral extraction on their planet.

The alternative hypothesis advanced by Tellinger is that close to a collision with Mars occurred on an earlier return of Niburu within the solar system, and this led to the loss of the atmosphere and water from Mars. (This does not, of course, explain the scientifically established relicts of former nuclear explosions on Mars!).

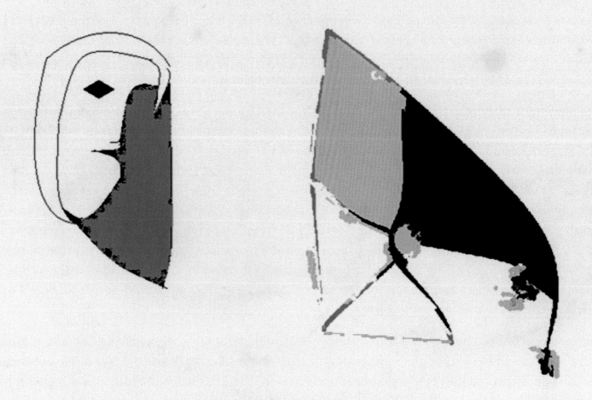

Figure 25. Drawing of the Cydonia 'Face' on Mars and an outline of one of the pyramids in the vicinity.

The hypothesis of Knight and Butler that super-intelligent beings from elsewhere might find it desirable to maximize the number of other intelligent species in the cosmos, contrasts with the usual idea that extraterrestrials come to conquer. It seems to the author that the assumption that advanced human beings coming from the future by time travel were needed to complete this task is unnecessary. Apart from the temporal contradictions this hypothesis implies, presumably it was based on the belief that we are the unique intelligent species in the galaxy, and/or that intelligent organisms only help their own kind? If Category II extraterrestrials were available, who by definition would be capable of modifying a solar system for whatever purpose, their involvement would represents a simpler hypothesis: one that is also compatible with the optimistic belief, namely that the overall cosmic intention is to fill the universe with intelligent species!

The frequent kidnappings of our species reported recently[16,18] (but which may have been going on for millennia) also suggest that although we have value as a labour force, we may be even more valuable as a source of genetic material. At the same time, we may have to modify the usual assumption of writers of science fiction that humans will eventually set out bravely on our rockets to conquer an uninhabited universe. We may just get a lift from friendly ET's!

Could it be that large numbers of our species who were 'sampled' in the past by UFOs are already on distant planets and space stations, employed as members of the ET workforce?

In contrast, the Sumerian account (presumably given them by the Anunnaki), refers to a glancing blow by the planet Nibiru while passing the Earth, resulting in Nibiru's expulsion from the solar system in an extreme orbit taking some 3,600 years to return periodically to the central solar system. This collision might also have given rise to the moon from displaced debris, although the Sumerian account suggests that the collision debris also contributed to the asteroid belt. A further hypothesis, suggested by Ingo Swann [78], is that the moon is older than the solar system, artificial, hollow, and inhabited by advanced species, who contest our assumption that it belongs to us. The orbit of Nibiru can be envisaged as similar to the comets which spend a great proportion of their long orbits outside the planets of the inner solar system.

Several intriguing questions arise from these largely hypothetical accounts:

- Is the hypothesis of a nuclear explosion leading to the loss of Mars' atmosphere tenable, or was it inevitable that Mars would lose its atmosphere and water bodies due to its low gravity and solar radiation, unmitigated by a strong magnetic field, such as exists on Earth?
- That there was once a high civilization on Mars at a very early date is supported by abundant artificial structures abandoned on the surface of Mars (Fig 25) and seen in NASA photographs. Were all these produced by a native population of Martians or much later by the 'Watchers' placed there by the Anunnaki?
- Why do these Martian structures, notably pyramids, resemble those of pre-Pharaonic civilizations on Earth? Which civilization built them? The native population, afterwards exterminated, or the Anunnaki Watchers? Was there a transfer of populations and culture between our two planets?
- Did the category II species detected by remote viewing while they attempted to mediate the Martian-Phaeton conflict, see any special significance for the future role of Earth as an incubator of intelligence? Is the Earth (as far as we know) the remaining inhabited planet in the solar system, capable of replacing two intelligent species lost to nuclear war - and can we be protected (largely from ourselves), so as to avoid a similar fate?

John F. Caddy Ph.D.

Is Panspermia a Normal Means of Spreading Life Forms (DNA) between Planets?

Crick and Orgel[5] supported the validity of the directed panspermia theory, namely that organisms were introduced onto Earth by intelligent beings from another planet, is more likely than the random distribution of spores through space. At the same time, random panspermia caused by material ejected after meteorite collisions on an adjacent planet seems a possible way that genetic material, notably DNA, could have arrived here (see cover drawing), given that bacteria and simple animals such as tardigrades have now been shown to survive in space. The other reason being that the long time needed to recreate the structure of this highly complex molecule by random processes is incompatible with the relatively short time from the age of the first fossils discovered in early rocks formed closest in time to the cooling of the Earth's surface. This again seems to favour panspermia.

One other implication of extended panspermia in transmitting DNA has not been drawn attention to: the possibility that certain configurations of living organisms may recur on a series of adjacent planets (e.g., the bipedal hominid pattern). Could this, for example, be a result of 'farming activities' by category II species in the distant past? Theories have been proposed that Martian meteorites could have transmitted DNA in this way, but as far as I know, organic debris from planet Phaeton (via the Asteroid Belt) has not been considered a feasible source—an unresolved but fascinating issue! The fact that planetary debris from planet Nibiru would also have carried DNA onto Earth (assuming life had already begun on that planet) was suggested by Evans as a possible reason for the genetic compatibility of their respective humanoid inhabitants, which allowed subsequent gene transplants between them. Or could this pattern be the result of farming activities by category II species on all planets of the solar system in the distant past?

A very recent event which can only be explained by directed panspermia, is the discovery in orbit around the Earth of microscopic metal spheres containing organic material.[73] Perhaps these spheres were manufactured in a way that they would open when they came down to Earth? During July 2013, Professor Milton Wainwright and his team sent a weather balloon almost 17 miles above the earth to collect samples. The most noteworthy capture was a metallic orb, 30 micrometres in diameter, composed of titanium and a tiny amount of vanadium. Testing revealed biological material oozing from inside the sphere, and on the outside a fungus-like covering. An observation supporting this discovery occurred in 1984 when the passengers of a train travelling through Padriesko, Moscow, witnessed a metallic craft landing nearby. Yuri Simiakov, a biologist and Russian ufologist, discovered not only elevated magnetic levels at the site and an absence of all living things—including microbes—but also tiny metallic spheres containing DNA. He quoted a team of Swiss and German scientists who had proven that DNA can survive the stresses of atmospheric re-entry. If intelligent beings wanted to transport biological matter to Earth from elsewhere, using small titanium-vanadium balls was an excellent choice. They are lightweight, resistant to high heat, and help shield their contents against solar radiation.

The theory of panspermia proposed by Wainwright and Wickramasinghe supposes that intelligent beings purposefully placed microorganisms within protective spheres to seed Earth. Perhaps these spheres were manufactured to open when they came down to Earth? The composition of one of the Wainwright spheres—titanium and vanadium—is shown to be custom-manufactured.

My friend Federico Bilotti, an engineer and expert on prostheses used in surgery, commented:

> *"Interesting to read about titanium found normally with nickel and aluminium in the mix used for surgical staples: they are made of 90 per cent titanium, 3–6 per cent aluminium, 1–2 per cent vanadium. (The spheres found in the stratosphere had 1 per cent vanadium).Thus alloy is used to obtain a movement when it is heated or traversed by electrical current when it is characterised by super elasticity and shape memory (the material remembers a shape, and when heated or electrically charged, it takes on the old shape). A liquid substance found inside could easily be introduced during the manufacture of the spheres in their original open state (two halves of a hollow sphere connected at their tangential circumference). On reaching a programmed temperature they close, and with the opposite temperature change, they could open, responding to their shape memory, as commonly employed in medical devices today".*

Of course there is no guarantee that the microspheres observed had this property, however useful it might be! Alternatively, when the bacteria reach a temperature at which they start metabolizing, would they create enough gaseous pressure to spew out of their casing? Panspermia theorists believe this could explain the sudden appearance of certain plagues and infectious diseases, but more radically in the context of this discussion they provide a mechanism for panspermia, assuming they were manufactured off-planet.

An optimistic view holds that visiting species of advanced intelligence may have had more altruistic motives that we do not yet understand—and, quite logically, peaceful behaviour could be a requirement for a species wishing to avoid extinction in the long run. Opinions have been expressed that a confederation of intelligent species in the local cosmos aim to protect us from aggressive exploitation by others until we emerge from our own phase of political barbarism. This may explain why reports on the deactivation of nuclear warheads by UFOs are common on the web (see, for example, www.collective-evolution.com). If we ever achieve an adequate stage of social maturity globally, we may be permitted to become a space-going species; otherwise, opinions have been expressed that it would be too dangerous to let loose on an already inhabited galaxy 6.5 billion aggressive astronauts! Assuming that category II civilizations are pacific in intent and seek to reduce the use of nuclear devices, in the long term, the survival of an intelligent species depends on it avoiding offending other intelligent species who may own super weapons of the planet-busting type. In addition to the diversion of meteors onto rebel planets, thermonuclear devices potentially used as tools for interplanetary conflicts, such as was suggested to have taken place between Mars and Phaeton.

Nuclear devices used as planet-busting tools are a major threat to planetary civilizations that can be appreciated by all intelligent species. However, a simpler method of extermination of a planetary population would be by diverting meteors to drop large rocks down the gravity well of a planet. Bearing this in mind should cool excessive aggressiveness by a newly space-born species!

A Personal Summary of Human Historical Landmarks

As a moderately well-read non-specialist, I was interested to see if a more or less coherent account of human prehistory could be assembled from the fragmentary material available, with the minimum of guesswork. You may judge whether I came close to my objective, or whether this book should be read as Science Fiction!

The Sumerians were the first civilization we know of who rapidly developed a successful urban life with irrigated agriculture and a high diversity of farmed food sources. The archaeological record does not show where they came from originally. Apart from detailed practical information on agricultural production and trade and commercial transactions, tablet inscriptions provided information on their 'gods', the Anunnaki— including their apparently physical origins, characteristics, and customs. The recording of information on clay tablets that were not biodegradable provided us with information in excess of that surviving from later civilizations. The translation of their tablets began in the late 1800s, but the reinterpretation of this information source owes an overwhelming debt to the lifetime work of Zecharia Sitchin. According to his reinterpretation, an extraterrestrial species referred to by the Sumerians as the Anunnaki, landed in Mesopotamia some 400,000 years ago and cultivated a civilization with world-spanning connections. They were described as large hominids resembling humans, with a paternalistic attitude to their later human subjects which led to our intellectual development and finally independence. Their priority was a search for gold, initially concentrated in the coastal seas of the Arabian Peninsula, but from roughly 200,000 years ago, there was a refocus on southern Africa and an extensive and numerous series of mysterious structures and associated terraces that were created on the southern African veldt.

Another unexplained discovery that seems to fit in with the Anunnaki interest in mining gold are the 'mega-statues' of faces left buried by the first central American civilization, the Olmecs. As noted by Tellinger, these relicts show facial features that resemble Africans, and the sculptures seem to portray experienced miners wearing protective headgear.[44] Were they 'expert miners' transplanted by the Anunnaki from southern African goldmines to mines in Central America? Megalithic sites such as those in pre-Pharaonic Egypt and Teotihuacan, Mexico, all share common features such as the high precision of the stonework and the use of massive blocks of hard rock held together by poured metal linkages: the stones are often aligned to cosmic or compass directions.

Several 'mythologies' or, perhaps more accurately, biblical accounts of events such as the Garden of Eden and Noah and the Great Flood, appeared first as very similar stories in Sumerian writings. There they describe factually the 'servitude' of human populations. In the book of Genesis of the Old Testament, the dominant personage in the Garden of Eden story is referred to as the universal Deity, but in the Sumerian writings it is an Anannuki noble, Enki. Together with the book of Enoch (who was the grandson of Noah), this echoes several accounts in earlier Sumerian writings which resemble biblical stories. According to Tellinger, humanity was guided into considering their guardians as deities, with all the consequences this had in obscuring our understanding of real events during the evolution of those world religions which share a common historical and geographical origin. In part, this explains the automatic assumption of early translators that the Anunnaki were 'sky deities'. It also reflects the fact that in the late 1800s, a description of a real physical entity descending from the sky was obviously a fantasy. This was not the case for Sitchin, given his knowledge while working, of the contemporary reality of ongoing space programs. The key question anyone reading Sitchin or books by his followers must ask themselves, is whether they believe that the rapid transformation of primitive hunter-gatherers into sophisticated city dwellers could have been feasible without external assistance? More fundamentally, they could ask whether it is reasonable to assume that civilizations survived through the extremes of an Ice Age without help from elsewhere?

The addition of an account of a very early 'civilization' in southern Africa before our reported history, and the subsequent migration northwards of 'escapees' to form the pre-Pharaonic civilization in Egypt, provides a convincing temporal coincidence between the pre-Sumerian myths and fragmentary anthropological information on human migrations out of Africa. It appears to suggest the idea of a controlling ET presence at this critical stage in the evolution of our species.

So what happened to the Anunnaki, given that they do not appear any more to be a dominant force on Earth? It seems improbable that they would have left our planet, given the problems they faced with the deteriorating climate of their planet Nibiru. A further speculation to end this structure of conjectures is that they decided to stay, and took over the only continent not then inhabited by hominids, Antarctica; before it was moved to its current polar position. Recent observations suggest the presence of pyramids there, and what have been called aliens with accompanying technological marvels. An ice-free Antarctica coast may be the location of the port of origin for those high sea vessels carrying tall, knowledgeable individuals in visits to post-Flood societies in reconstruction—visits speculated on by Graham Hancock with respect to Egypt, Mexico, and Peru. Might these visitors even have been refugee emissaries from the sudden 'refrigeration' of their civilization on Antarctica?

The Anunnaki insistence on not passing on to our species their greater lifespan was certainly not their greatest gift to humanity. However, a short lifespan with no workable plan to stem our exponential rise in population could explain why the Anunnaki did not persist in taking over our planet. It must have been clear to them that their species could not compete with a short-lived, highly fecund species engaged in a population explosion which took us from perhaps a few million in 2,000 BC to 7 billion four thousand years later! Without implementing a policy of human extermination or active population control, which would have been contrary to their ethics, the Anunnaki would soon have been seriously outnumbered. The only strategy available to the former master species may have been to control the human population by subterfuge (i.e., the 'Tower of Babel' strategy—establishing geographically separate and eventually culturally and linguistically distinct human communities, who would control their own populations by mutual aggression). We now face the so-far insoluble problem of how do *we* control our own populations in a world facing a shortage of food and space, chemical contamination and a dangerous excess of atmospheric carbon compounds from mass combustion! From this perspective then, a 'reinterpretation' of human history as to our origins and the decisions which led to the current unsafe situation, could still be useful.

A related problem is how to account for the disappearance of the Megalithic societies, given their technological superiority? Did they leave, or was the explanation suggested by Sitchin valid: a relative sterility of hybrids between our two species? It would be useful to have evidence for infant Anunnaki-human hybrids in archaeological investigations, but few of these have been reported.

Our Future in the Cosmos

The remote viewing exercise just mentioned was where human 'disincarnate' observers paid a 'remote visit' to two protagonists, the Martians and the Reptilians, before nuclear war effectively destroyed their planets or rendered them uninhabitable. Our observers from their future were aware of a representative of a higher civilization present who was attempting to defuse the conflict. Assuming that this highly evolved species is still concerned to replace two intelligent species lost from the solar system, this could explain why the preparation of the earth for the evolution of our intelligent species should have had some relevance to them. One can only hope that such superior entities in the galaxy still look favourably on us and are prepared to guide us out of the geopolitical conflicts we are always creating.

New techniques based on a better understanding of human sensory capabilities may eventually allow us to access mental records made by multiple intelligent species stored in the hypothetical cosmic 'memory banks' postulated by various experts. My personal view on spiritual matters in the cosmos coincides closely with those described by individuals who have contacted extraterrestrials such as Steven Greer, and also resembles a view expressed as a theory of physics by David Bohm.[67] Bohm's concept of the Implicate Order is a hidden system of rules diffused throughout space, which regenerate the universe microsecond by microsecond. A cosmos-wide Mind which registers the thoughts of all intelligent creatures throughout the Universe is not far removed from this idea. If approached in a meditative state, this cosmic over-mind may provide inspiration and information to humans on past events and discoveries by higher species of relevance to ourselves.

Concerning their knowledge of practical astronomy, many ancient peoples knew details about the planets in the solar system that we only rediscovered in the last century or so. It is not obvious where this information could have come from, given that they did not have access to high technology—unless they had received this knowledge from a technologically advanced species. An Akkadian cylinder from the third millennium BC based on Sumerian precedents, shows a group of eleven circular objects encircling a rayed star. Sitchin's interpretation of these as individual planets, was facilitated by their relative size and distance from the sun (Fig 9). Included there is a relatively large unnamed planet called Nibiru, with an extreme trajectory which results in this major planetary body entering the central solar system between Mars and Jupiter at intervals of 3,600 earth years.

That in 4,500 BC the Sumerians knew there were planets beyond Saturn, a fact discovered by us within the last one hundred years, presents a conundrum which can only be solved by the presence among them of a space-travelling race. One other observation that emphasizes this point is that the Sumerians referred to the Earth as the 'seventh planet'. Only by counting the planets starting from the farthest from the sun is this nomenclature correct, and this would be logical for a species that at that time was living outside Pluto.

When discussing the possibilities implicit in the presence of a more advanced species living among us in our early history, we should consider the quotation that to primitive peoples (ourselves), advanced science appears like magic. Another useful axiom is that an event that is highly improbable on an annual basis or during a short lifespan, is almost certain to occur over a geological time frame and during a longer lifespan! Thus what we call logical reasoning for a short-lived species may be deficient when we attempt to reconstruct our origins incorporating information from long-lived extraterrestrials and guessing at their motives. Imaginative hypotheses such as this one are inevitably wrong in details but should be judged by the number of unexplained phenomena which they do *not* account or. If you found this story unconvincing, as said before, please feel free to read this text as science fiction; my objective was mainly to see what could be assembled from existing information to provide a coherent story that can be modified by further discoveries. After all, many experts predict that elements from sci-fi ended up as reality a few decades later! Therefore, we should soon expect to meet entities who command science and technology far in advance of our own. Perhaps they have information on our early origins that is needed to correct this report?

If we are fortunate, these civilizations could be of category II—a classification referring to their potential capacity to modify a solar system for their own benefit (or for others, as in the hypothesized case of the Earth). An (optimistic?) feature of advanced civilizations might be their desire to encourage the evolution of intelligence in the cosmos, and such an altruistic motive does not seem improbable.

If this hypotheses as to how cosmic visitors may have impacted our evolution has some validity, we should try to identify what actions we should then take. Evidently, these actions would not necessarily be different from those we should be implementing for our own good if we had evolved independently from cosmic interventions. For example, tackling the causes of climate change, pollution, wars, and in particular, aiming for the non-use of atomic weapons, are top priorities we do not take seriously enough. I suggest that this is because we hope that disastrous events will only occur after our short lives are over.

Three distinct paths into the future have emerged from recent advanced research programs that deserve priority attention. These are, respectively, that:

1. Within no more than twenty-five years,[76] work on artificial intelligence is predicted to result in an artificial thinking mechanism with a performance beyond human mental capacity and a functional duration greatly exceeding our lifespan. Should we turn over the decisions needed in a complex society to this advanced computing entity? A related question is whether off-planet expeditions should be turned over to robots with superior artificial intelligence that are immune to the problems humans experience in space?

2. Should our species allow ourselves to evolve according to the slow and largely unknown impacts of Darwin's 'natural selection' on city dwellers, or seek a faster and more efficient strategy? Advances in genetic research, such as the CRISPR procedure have now placed in our hands the ability to modify our genome artificially. We can agree, I suppose, that eliminating hereditary diseases is a major positive of this technology, but do we also accept that those with extra financial resources

should be able to arrange that their progeny, or they themselves, can evolve greater longevity, intelligence, and telepathic capability — all eventually leading to reinforcing an effective caste system in our society? Should there be moral criteria to decide on who has access to these costly transformations? Genetic improvements using these procedures may nonetheless be necessary to create a human genome better adapted to safer space exploration or even to counter the more extreme environmental conditions on planet Earth once climate change has occurred.

3. The motive force driving UFO's is said to be energy extracted from the vacuum. Adopting this producing energy technology would of course rapidly reduce CO_2 production and help solve global warming, but its research and adoption risks being blocked by giant corporations dependant for their profits on carbon-based combustion, since such a revolution could be implemented by small local industries and become independent of industrial outputs.

The impression gained from interventions by our 'Visitors' suggests that they see continued human evolution as a priority, compared with humans relying entirely on technological advancement. The idea of continued human evolution over the long term is implied by the cover diagram at the beginning of this text. Panspermia ensured an early start to the evolutionary process on the earth's surface with Darwinian evolution as the main process moving life to more complex forms and functions. The mystery is that for intelligent life forms there seems to have been a transition to what may be called a 'syntropic process' (see, e.g., Ulysse Di Corpo[80]), where the final result is a call from the future that appears to draw us onwards. One result of this positive factor has been a rapid evolution in intelligence, speeded up by an unnamed 'attractor' which is drawing us inexorably to the exploration of space to encounter and learn from other intelligent species who arrived there much before us. This seems to be leading us to becoming an organism with the capacity to live in a gravity well as before, but we can also be free of the planetary surface in our flight through an infinite universe. As a life form with superior intelligence, we will be capable of deciding on our future evolutionary trajectory — supported by our inventions but not dominated by them.

Some Tentative Conclusions

This text has described a personal investigation into the the literature on the early history of humanity by a person who is already convinced of the existence of extraterrestrial civilizations in our galaxy and beyond. For my own purposes as a scientist, I am concerned to know how this possible prehistorical trajectory for our species—let us say from earlier than 4,000–5,000 years BC—might have been realized. In an attempt to interpret prehistorical information new methodologies have now come into play, notably genetics taking us back to our origins as a species; and cosmic information, resulting from our investigations of the solar system. These sources provide information we have to incorporate into any study of human prehistory.

Sitchin's main conclusion is that we owe our rapid rise to intelligence after 12,000 BC or so to an extraterrestrial species. This milestone has not been widely accepted, although many of the Sumerian inscriptions he brought to light remain difficult to interpret otherwise. We are not in a position to make definitive statements on many issues of importance to this debate, but to my mind, the key to making progress is to erect an overall hypothesis that incorporates many individual discoveries within it. The following elements are selected for personal emphasis:

1. Starting with the cause of the origin of life on Earth, there seems a good case for assuming panspermia was a driving force, given the rapid onset of life forms following the cooling of the early earth's surface. The recent discovery of metal microspheres in the stratosphere containing organic material makes the hypothesis of 'directed panspermia' very plausible.

2. That the Anunnaki were extraterrestrials who arrived on this planet roughly half a million years ago seems to be confirmed by (a) a number of Sumerian inscriptions showing designs of rockets and spaceship landing sites and (b) accounts of key events also documented some thousand years later in Genesis and the book of Enoch.

3. The characterization by the Sumerians of the Anunnaki as divine, reflects a primitive human tendency to assign this characteristic to strange powerful beings coming from elsewhere, and this belief was apparently encouraged by their masters, who came from a rigid theocracy. That the followers of the first translators of Sumerian in the nineteenth century were scandalized by Sitchin's conversion of the texts from describing activities of mythical/divine personages to descriptions of reality is not surprising, given that space flight was not yet a reality at the time of this first translation.

4. By biblical times, some thousand years or so later, the stories first recounted by the Sumerian scribes such as the event in the Garden of Eden, seem linked in meaning to the biblical account of the 'awakening' of Adam and Eve to their spiritual responsibility. We may now postulate that this upgrading of our human perception followed from genetic improvements to our mental capacity and from the education we received, which allowed early hominids to perform useful work for the Anunnaki. In the original story the 'serpent' was not necessarily evil, but perhaps in this allegory reflected a religion of the Anunnaki descendants which was described in early texts as the Serpent Cult, where serpents were perceived as wise organisms. Perhaps this, in turn, was a memory of an earlier reptilian connection for pre-humans?

5. In South and Central America, Mesopotamia and Egypt, there are close similarities in Megalithic construction techniques which suggest common worldwide themes. These nonetheless involve real mysteries in the technologies used. They suggest that at the time of construction there was a worldwide civilization promoted by the Anunnaki and their descendants using methods developed off-world.

6. The ability to move and precisely shape hard rocks weighing up to a thousand tons in some cases, then to transport them hundreds of kilometres from the quarry — is still unexplained by science. They differed from the usual rectangular shaped blocks used in human constructions, being cut precisely in irregular geometrical shapes that nonetheless fitted together perfectly. Another common theme linking megalithic sites around the world are pyramids, which apparently were not primarily intended as burial sites. Were they perhaps used for signalling or energy generation? There seems evidence here for technological applications unknown to modern science, such as the use of sound energy, pranic energy or antigravity.

7. Large engraved images on the ground in South America, Asia, Antarctica, and on Mars, suggest that these were 'signposts' used to identify the location of Anunnaki communities to space pilots. Statues of visiting 'gods' found in Bolivia, Mexico, and Turkey, unlike South American peoples who are generally without facial hair, portray them as bearded and either clothed in fish scales (said to be characteristic of Anunnaki space pilots). They were carrying a characteristic 'handbag' similar to modern pilots when outside their vehicle (Fig 15). Most of them are also wearing something resembling a time piece on their wrist. The similar appearance of Anunnaki-like 'gods' in these widely separated locations suggest the Anunnaki arrived there by air or space travel.

8. The presence of a higher civilization prior to the onset of the first documented human civilizations in Sumer, Egypt, Turkey, Peru-Bolivia, and the Indus Valley, has been speculated on for a number of reasons by Hancock, and Knight and Butler. The reason for considering that there was an off-planet basis for these early civilizations was their rapidity of development, with urban centres supported by commercial agriculture, trade, and flourishing artisanal production. More surprising still was the development of a focus on studies requiring long-term spatiotemporal astronomical observations in the Sumerian, Egyptian, and proto-Mayan cultures, with a necessary duration longer than that recorded for the civilizations themselves.

9. The reproductive experiments carried out by the Anunnaki on primitive humans are described in Sumerian texts and by simple drawings, reflecting probably a lack of knowledge by the scribes of the theory underlying the introduction of Anunnaki genes into the human genome. As might be expected, the first attempts at gene transplants were reported as rarely successful, and the fertility of the first progeny was low. The low diversity of male chromosomes is reflected in the

general uniformity of modern Y chromosomes. This could be due to the low number of ancestral males used to produce the sperm employed in artificial fertilization, or the subsequent 'cultural cleansing' of males in subjugated hunter-gatherer peoples during the geographical expansion of humans upgraded by the Anunnaki; a feature of our species behaviour also seen in more recent historical events.

10. A hypothesis is put forward based on the work of Julian Jaynes which throws some light on the psychology of ancient Mesopotamian peoples from Sumerian times onwards, after their genetic modification by the Anunnaki. His conclusion was that their mind set was essentially pre-conscious (bicameral), in being vulnerable to hypnotic suggestions through voices heard from the silent hemisphere of the brain. This voice repeated recent instructions given by the priest in temples dedicated to the king of the Anunnaki, who was visualized as a divine being. Such a hypothesized mechanism suggests how left brain dominance came into being under the Anunnaki regime, which for the human slaves was absolutely totalitarian. Historical evidence for the loss of human rights under more recent totalitarian regimes seems to reflect the Anunnaki influence.

11. Direct hybridization was also carried out, including cross-fertilization of human females by Anunnaki males (and perhaps vice versa), and this produced quite different offspring: the giant cone-head humanoids, a hybrid race referred to as the Nephilim in the Bible. These Anunnaki-human hybrids were apparently the rulers of megalithic civilizations including pre-Pharaonic dynasties in Egypt.

12. Michael Tellinger describes the discovery of the extensive ruins of what may be the most ancient city on the planet in southern Africa, provisionally dated to some 200,000 years ago. This gives additional credibility to the Sumerian account. Whether these buildings were dwellings or closed compounds to house human slaves working in adjacent goldmines, is still to be ascertained. The additional confirmation of Sitchin's hypothesis of the linkage of this southern African to the early Egyptian civilization comes from the discovery of proto-Egyptian carvings there, suggesting that the subsequent migration northwards of 'escapees' could have provided an impetus to the founding of the pre-Pharaonic civilization. This is usually considered to have occurred in a mythical period at the end of the last Ice Age some 12,000 years ago, often referred to as the age of the 'Zep Tepi'.

13. That the Anunnaki or Nephilim were involved as early rulers of Egypt in this period seems likely from tombs of giant people found in Northern Egypt and worldwide. Their high technologies, in the form of pyramids and megalithic stonework, was apparently developed in a different direction from recent human science, suggesting there is still much to be discovered!

14. There appeared to have been a brief resurgence of control by the Anunnaki line when Amenhotep, a pharaoh physically similar to a cone-headed hominid, came to the throne in Egypt. This event was important for the eventual transition of global societies from polytheism to monotheism,

but what was the cultural connection for this drastic change? It is difficult to believe that this was entirely his personal inspiration. Did it reflect Anunnaki beliefs which saw divine beings associated with each planet of the solar system?

15. A well-documented series of discoveries of skeletons of so-called cone-headed humans of more than average human size have been discovered around the world, often associated with megalithic constructions. These do not fit in with the orthodox evolutionary history of our species. Their much larger brain size and the lack of key skull sutures exclude the mechanism often quoted: that this was due to a technique used on these giants to influence their head shape as infants by compressing the flexible infant skull by 'head boards'. This practice of compressing the skulls of human babies to make them conform to an artificial 'cone-head' shape, could in essence have been inherited from a time when this practice was used to emulate 'cone-headed' beings who were their social superiors in mixed species communities.

16. The existence of a prenatal cone-head fossil confirmed their separate identity from humans, while preliminary genetic tests on an adult skull sample from Peru identified it as a hybrid between a Caucasian father and a non-human mother. That these hybrids were long-lived compared with humans, and at least occasionally infertile, may help account for their apparent disappearance from the historical record after 3,200 BC (the date ascribed to the seven hundred Nephilim skulls found in a Maltese temple).

17. The apparent disappearance of the Anunnaki from Earth is a mystery, considering the problems being encountered with heat loss on their home planet Nibiru. A possible answer to this question could be that they occupied Antarctica before its catastrophic move to the South Pole which has been postulated to have been caused by a major crustal displacement. This could in turn, have resulted from a shift in the poles of rotation of the Earth. Such an event may also have been responsible for eliminating the mega-mammalian fauna of North America and the mammoths in Siberia. The Anunnaki who survived this event, and perhaps also their hybrids, may then have decided that the Earth was too unstable for a very long-lived species and returned home? The presence of ground images left by the Sumerians, ancient Egyptians, and Sri Lankans in Antarctica, is indirect evidence of ancient trade routes, including Egyptian hieroglyphics found in Australia. The recent discovery of ancient artificial structures, cities and pyramids, crushed under kilometres of ice in Antarctica may offer still more surprises for future historians.

18. So where does this painful account of our training under an extraterrestrial Intelligence leave us? Perhaps a practical conclusion from this study is that we should avoid automatically accepting commands from higher strata in Anunnaki-inspired totalitarian systems originally designed to incorporate slave labour. We should discuss the orders received from above with our colleagues,

clan members or co-workers, as would have been appropriate in our original hunter-gatherer communities, and if necessary, request clarification from above.

19. Searching for an overall hypothesis that brings the cosmic phenomena discussed here to an overall conclusion is the predominance of the hominid pattern in the ET species who have visited the Earth, and the possibility of hybridization. Hence some commonality in their DNA with terrestrial hominids may be a significant element.

20. Once this ancestral bipedal species had invented a means of cosmic travel, they voyaged through space in a search of those scarce worlds having a basis of liquid water but no life forms. Perhaps they introduced DNA-based forms by accident or design onto these lifeless worlds? And perhaps their long-term intent was to use panspermia to create suitable worlds for their future expansion? However, over a much longer evolutionary term perhaps this action may have spread intelligent hominids throughout the cosmos!

References

(1) Sitchin, Zecharia. (1976). *The Twelfth Planet*. HarperCollins Publisher, New York.

(2) Knight, Christopher, and Alan Butler. (2004). *Civilization One*. Watkins Publishing, London.

(3) Knight, Christopher, and Alan Butler. (2005). *Who Built the Moon?* Watkins Publishing, London, 262p.

(4) Hancock, Graham. (1995). *Fingerprints of the Gods*. Three Rivers Press, N.Y.

(5) Crick, F., and H. Orgel, L. E. (1973). "Directed panspermia". *Icarus* 19 (3): 341–346.

(6) Fenton, D. (2018) 'Hybrid Humans: Scientific Evidence of Our 800,000-Year Old Alien Legacy'. www.amazon.com.

(7) Sitchin, Z. (2004). Official website of Zecharia Sitchin. On the trail of planet X. (http://www.sitchin.com/frenchastron.htm)

(8) Schmidt, G. A., and A. Frank. (2018). "The Silurian Hypothesis: Would It Be Possible to Detect an Industrial Civilization in the Geological Record?" *Int. J. Astrobiology*, pp. 1–9, 2018. (http://dx.doi.org/10.1017/S1473550418000095).

(9) Prashanth, Damara. (2017). Bottom of Form Researchers Find Three-fingered Alien Mummy in Nazca, Peru: Facts. Web journal: Paranormal (Hoax or Fact).

(10) Tellinger, M. (2015). South Africa's Ancient Annunaki Goldmines. *Environment*, 14 May 2015.

(11) Tellinger, Michael. (2005). *Slave Species of the Gods: the Secret History of the Anunnaki and Their Mission on Earth*. Bear and Company, City.

(12) Scott Jones, C. B., and Angela T. Smith. (2014). *Voices from the Cosmos*. Headline Books Inc., City.

(13) Evans, M. J. (2016). *Zecharia Sitchin and the Extraterrestrial Origins of Humanity*. Bear and Co., Vermont, USA.

(14) Walla, Arjun 2017. Astonishing Updates on the Potential Alien Body Unearthed In Nazca, Peru. Unlike Anything We've Ever Found. In: CE Collective Evolution, 15/09/2017. https://www.collective-evolution.com

(15) Nalini, Polavarapu; Gaurav Arora; Vinay K Mittal; John F McDonald. (2011). Characterization and Potential Functional Significance of Human Chimpanzee Large INDEL Variation. *Mobile DNA* 2(13), DOI: 10.1186/1759-8753-2-13.

(16) Robinson, S. P. (1998). 'UFOs, Aliens, and Alien Abductions'. http://www.niu.edu/ newsplace/ nnaliens.html.

(17) Bloxham, A. (2010). 'Aliens Have Deactivated British and US Nuclear Missiles, Say US Military Pilots'. *The Telegraph*: https://www.telegraph.co.uk/news/newstopics/howaboutthat/ufo/8026971/Aliens-have-deactivated-British-and-US-nuclear-missiles/ Sept.2010.

(18) Shcherbak and Makarov. (2013). 'The "Wow! signal" of the terrestrial genetic code'. *ScienceDirect. Icarus* 224(1), May 2013, pp. 228–242.

(19) Jha, Alok. (2004). 'Human Brain Result of "Extraordinarily Fast" Evolution'. *The Guardian* (29 Dec 2004). https://www.theguardian.com/science/2004

(20) Vanderhaegen, P.. (2018). 'How Did Human Brains Get So Large?' European Research Council, ERC (https://erc.europa.eu/projects/figures/stories).

(21) Tattersall, I. (2004). 'The Dual Origin of Modern Humanity'. *Coll. Antropol* 2004; 28 Suppl 2:77–85.

(22) Alexei A. Sharov, A. A., and R. Gordon. (2013). 'Life before Earth'. Cornell Univ. Library. (arXiv.org > physics > arXiv:1304.3381).

(23) McBrearty, S., and A. S. Brooks. (2000). 'The Revolution That Wasn't: A New Interpretation of the Origin of Modern Human Behavior'. *Journal of Human Evolution* 39; pp. 453–463.

(24) Top of Form MacIsaac, T. (2014). 'Did Ancient People Really Have Lifespans Longer than 200 Years?' *Ancient Origins*, 21 September 2014.

(25) Gaia staff 2018 Powerful Evidence of Nuclear Wars in Ancient Times https://www.Gaia.com/article/April 19 2018

(26) Yirka, Bob. (2013). 'Researchers Use Moore's Law to Calculate That Life Began before Earth Existed'. https://phys.org/news/2013-04-law-life began-earth.html. 18 April 2013.

(27) Pennisi, E. (2014). 'Human Speech Gene Can Speed Learning in Mice'. *Science*, September 2014.

(28) Wikipedia, 'Cargo Cult'. https://en.wikipedia.org/wiki/Cargo_cult.

(29) Finkel, Irving. (2014). *The Ark Before Noah: Decoding the Story of the Flood.* Hodder Books, City.

(30) 'Akhenaten: King of Egypt'. *Britannia.com.* **https://www.britannica.com**/biography/ Akhenaten.

(31) https://www.theancientaliens.com.

(32) Thom, A. (1962). 'Megalithic Sites in Britain'. (http://www.spirasolaris.ca/sbb8a.pdf).

(33) Hapgood, C. H., and James H. Campbell. (1958). *Earth's Shifting Crust.* Later revised and republished in 1970 as *The Path of the Pole*, by Chilton.

(34) Veall, W. J. (2017). 'Tamils and Sumerians. Among the First to Reach Australia and Antarctica?' Part I. *Ancient Origins.* 6 Dec. 2017.

(35) Anon. Full text of '**The Epic of Gilgamesh**', https: Archive.org/Stream/TheEpicofGilgamesh.

(36) 'Alessandro Morbidelli and the origin of Sedna'. (www.darkstar1.co.uk/Morbidelli.html).

(37) Djonis, C. (2016). 'Planet X—Is There Scientific Evidence?' *Ancient Origins*, April 2016.

(38) Bonanno, A., C. Malone, D. Trump, S. Stoddart, T. Gouder. (2005). 'The Death Cults of Prehistoric Malta'. *Scientific American*, 1 January 2005.

(39) Wikipedia, 'Book of Enoch'. https://en.wikipedia.org/wiki/Book_of_Enoch.

(40) Anon. (2018). 'Titans under the Earth: Evidence for the Tall Ones and the Ancient Mounds of Pennsylvania'. *Arjun Walia*, 8 JANUARY, 2018

(41) Griffiths, S. 2015.Strange'conehead' skeleton unearthed at Russia's Stonehenge.Mail Online, https://www.daily-mail.co.uk/sciencetech /Article-3176210

(42) Hancock, G. 'Megalithic Origins: Ancient Connections between Göbekli Tepe and Peru'. (https://grahamhancock.com/newmanh2).

(43) Ivan.2017. 'The Mystery Handbag of the Gods: Depicted in Sumer, America, and Göbekli Tepe' (Ivan is editor-in-chief at Ancient-code.com, where he published this comment in Ancient Code: 05/04/2017.

(44) Wikipedia, 'Olmec Colossal Heads'. En.wikipedia.org/wiki/Olmec-colossal-heads)

(45) Hall, S. (2012). 'Hidden Treasures in Junk DNA'. *Sci. American*, 1 October 2012.

(46) Anon. (2017). 'Update on the Discovery of an "Alien Body" Unearthed in Nazca, Peru'. *https://www.disclose.tv/update-on-the-discovery-of-an-alien- body-unearthed-in-nazca-peru-315388*Sep 16, 2017.

(47) Wikipedia, 'Dyson Sphere'. *https://en.wikipedia.org/wiki/Dyson_sphere.*

(48) Cremo, M., and R. Thompson. (1993). *Forbidden Archeology: The Hidden History of the Human Race.*

(49) Zimmerman, Fritz. (2010). *Nephilim Chronicles: Fallen Angels in the Ohio Valley.* Goodreads. https.//www.goodreads.com/book/show 8968478

(50) Gigal. (2018). 'A Giant Pharaoh Was Discovered According to Science?' htpps://www.gigalinsights.

(51) Anon. 'Who Were the Ancient Giants with Six Fingers and Double Rows of Teeth?' http://locklip.com/who-were-the-ancient-giants-with-six-fingers-and-double-rows-of-teeth/.

(52) https://www.express.co.uk/news/weird/694326 /Are-there-alien-skulls-New-DNA-tests-on-Elongated-Paracas-Skulls-could-change-history.

(53) Steiger, Rod. (2007). 'Exists Numerous Evidence of Pre-historic Nuclear War.' *The Canadian* website. www.bibliotecapleyades.net/ancientatomicwar /ancientatomic8.htm.

(54) Forti, K. J. (2015). 'Evidence of Ancient Civilization Nuclear Wars', 8 October 2015/2 (Comments/ in Blog). Trinfinity- htpps://www.trinfinity8.com

(55) Giantology.(2015). 'The Last 100 Years'. rephaim23, *Just another WordPress.com site.* Posted on 20 June 2015.

(56) Filer, G. (2012). 'Mars Tubes'. National UFO Center, 27 Sep. 2005. http://nationalufocenter.com/artman/publish/article_3php

(57) Anon. (2017). 'Strange Structures on Mars Connected to the Sphinx and Pyramids of Gizeh?' 8 July 2017. *Disclose TV*. Http.//www.disclose.tv/ July 8 2017

(58) Caddy, J. F. (2017). 'From Shamanism to the Space Age: Reconciling with Ancient Beliefs May Prepare Us for Contact'. *IOSR Journal of Humanities and Social Science (IOSR-JHSS)* 22(4) Ver. 5 (April 2017) pp. 58–71.

(59) Hoagland R. C. (2014). Chinese Lunar Images what did they discover. *Coast to Coast.* 22 April 2014. https://www.youtube.com/

(60) Brown, Courtney. (2011). 'The Origin of the Asteroid Belt: The Exploding Planet Hypothesis'. Presented at the 30th Annual Meeting of the Society for Scientific Exploration in Boulder, Colorado, 11 May 2011. In videos of recent talks.

(61) 'The Mystery of Abydos and the Osirion Temple'. *Ancient Aliens.* https://ancientaliens/wordpress.com/2011/01/04/the-mystery-of-abydos-and-the-osirion-temple/.

(62) 'Alien Technology: Quimbaya (Tolima) Airplanes'. *The Ancient Aliens.* https://www.theancientaliens.com/

(63) Little, G. (2014). 'The Truth about Giant Skeletons in American Indian Mounds, and the Smithsonian Cover-Up'. *AP Magazine*, 28 Jun. 2014. https://www.worldnewsdailyreport.com/

(64) The Farsight Institute. (2014).*The Great Pyramid of Giza: The Mystery Solved.* http://farsight.org/Farsightpress/

(65) Ivan. Megaliths of Baalbek: A Colossal Mystery of stones weighing over 1600 tons. *Ancient Code.* http://www.ancient-code.com/

(66) Wikipedia. 'Mesoamerican Long Count Calendar'. http://en.wikipedia.com /wiki/

(67) Bohm, D. (1980). *Wholeness and the Implicate Order.* Routledge Classics. London, New York.

(68) Wicksell, D. (2014). Physicist Claims Evidence Ancient Nuclear Explosions Ended Life on Mars. *Inquisitr*, 21 November 2014. https://www.inquisitr/1625693/

(69) Gaia Staff. (2018). 'Powerful Evidence of Nuclear Wars in Ancient Times'. *Gaia.* 18 April 2018.

(70) Winters, Clyde A. (1998). 'The Decipherment of the Fuente Magna Bowl'. http://olmec98.net/Fuente.htm.

(71) Steenstrup, T., and 28 co-authors. (2017). 'Telomeres and the Natural Lifespan Limit in Humans'. *Aging* 9(4): 1130–1140.

(72) Tian Chen Zeng, et al. (2018). 'Cultural Hitchhiking and Competition between Patrilineal Kin Groups Explain the Post-Neolithic Y-Chromosome Bottleneck, *Nature Communications*. DOI: 10.1038/s41467-018-04375-6. (Read more at: Https://Phys.org/News/2018-05-Wars Clan Strange Biological Event.html).

(73) McClure, K., and C. Kloetzke. (2017). 'Is Earth Being Seeded with DNA-Filled Metallic Spheres from Space?' https://thefieldreportscom.wordpress.com /2017/08/29 / is-earth-being-seeded-with-dna-filled-metallic-spheres-from-space/.

(74) Brown, Courtney. (2005). *The Science and Theory of Non-Physical Perception*. Farsight Press, Atlanta.

(75) Kurzweil, R. (2005). *The Singularity Is Near: When Humans Transcend*. Viking Press, New York.

(76) HHMI. (2004) 'Human Brain Evolution Was a 'Special Event'. Howard Hughes Medical Institute, *RESEARCH*, 29 Dec 2004.

(77) OSU. (2018). 'What's the Most Recent Eruption of Vesuvius and Will It Erupt Again?' Volcano World: Oregon State University.

(78) Swann, Ingo. (1998). *Penetration: The Question of Extraterrestrial and Human Telepathy*. CrossroadPress., USA.

(79) Barras, Colin. (2018). 'Survival of the Tamest'. *NewScientist*, 24 February 2018.

(80) Di Corpo, Ulisse. (2018). *Sintropia: La Trilogia*. Kindle Format.

(81) Anon. 'Coral Castle: How A Sick Man Used Ancient Wisdom to Build a Modern Wonder'. Ancientexplorers.com/blogs/news.

(82) Marzulli, L.A. (2016). 'New DNA Testing on 2,000-Year-Old Elongated Paracas Skulls Changes Known History'. *Ancient Origins*, 23 July 2016.

(83) Luskin, Casey. (2015). 'The Octopus Genome: Not "Alien" but Still a Big Problem for Darwinism.' *Evolution News and Science Today*, 24 August 2015.

(84) Caddy, J. F. (2018). 'Were Some Astounding Truths Left for Us By Our Cosmic Tutors?' *Syntropy Journal* 2018(1).

(85) Cossins, D. (2018). 'Farewell to Kepler, Finder of Worlds'. *New Scientist*, 10 Nov. 2018.

(86) Lovett, Richard A. (2017). The Invisible Hand. Could a bizarre hidden planet be manipulating the solar system? NewScientist, No.3156, 16 December.

(87) Jaynes, Julian. (1976). The origin of consciousness in the breakdown of the bicameral mind. Houghton Mifflin, Boston.

(88) Snyder, Michael (2014). Newly Found Megalithic Ruins In Russia Contain The Largest Blocks Of Stone Ever Discovered. INFOWARS Newsletter, Https//www.infowars.com, March 11, 2014.

(89) Planet 9. Wikepedia

(90) Full text of "Lost Book o Enki.pdf (PDFy mirror)" . Internet Archive. https://archive.org/stream/

(91) Anon. Ancient Egyptian mural rocket launching. https://uk.images.search.images.yahoo.com/

(92) Austin, J. 2016.The shocking claim mystery skulls 'found in Antarctica could be from aliens' JULY 26 2016 https://www.express.com/news/weird/693091/

(93) Climate change: past and future. In: Discovering Antarctica. British Antarctic Survey/Foreign and Commonwealth Office/Royal Geographical Society. https://discoveringantarctica.org.uk/

(94) Forgione, A. 2011. Malta's long-headed skull. In UNMYST3. (https://www.unmyst3.com/2011/04/), April 16, 2011

(95) Pasulka, D.W. 2019. How the Increasing Belief in Extraterrestrials Inspires Our Real World, VICE Magazine, Truth and Lies issue, Mar 11 2019.

(96) Holloway, A. 2014. Largest known megalithic block from antiquity revealed at Baalbek. Ancient Origins. December 2014.

(97) Natasha, P. 2018. Researchers discover early medieval women with their skulls altered in Germany. History and tech news. (htpps://www.histecho.com), 28th November 2018.

(98) Anon. Category: Media in categoria: Art of Mohenjo-daro. Wikimedia Commons. https://commons/wikimedia.org/

(99) Hatcher Childress, D. 2000.The evidence for ancient atomic warfare. Nexus Magazine Vol 7(5), Nov – Dec 2000, USA.

(100) Newman, H. 2016.Top ten giant discoveries in North America, In: Ancient Origins, 18 Jan, 2016.

(101) Olson D. 2017. 10 Photos that Prove Megalithic Engineers Predated the Inca Builders. November 13, 2017.

(102) Sutherland, A. 2015. Mysterious Ancient Rulers With Elongated Skulls – Who Were They Really? AncientPages.com, May 19, 2015.

(103) Keys, D. 2019. Neolithic Britons travelled across country for regular mass national feasts 4,500 years ago, new research claims. The Independent, Wednesday 13 March 2019.

Annex Table: A Hypothesized Chronology

Some critical events in the history of the human species before the Great Flood, our hypothesized origins, and their tentative chronology, based on the authors specified.

Events	Dates (yr)	Sources
Formation of Earth	4.54 billion years ago	Wikipedia
Start of life on Earth	3.5 billion years ago	Wikipedia
Mega-impact of meteor/comet destroys the dinosaurs	65 million years ago	Wikipedia
Mindel glaciation	424,000–278,000 yrs ago	Wikipedia
Anunnaki arrive on Earth	445,000 years ago	Sitchin
Homo erectus	1.6–0.4 million yrs ago	Hancock
Homo sapiens	400,000+ years ago	Hancock
Anunnaki in Southern Africa for gold mining. Global warming makes mine work unbearable for Anunnaki—on strike	285,000 years ago	Tellinger
In S. Africa, 200K–10 million ancient structures built	285,000 years ago	Tellinger
1st man/1st woman (genetic extrapolation)	170,000 years ago	Tellinger
Loss of Mars atmosphere + water due to Nibiru passing near Mars? Or caused by a nuclear attack?	155,000	Tellinger
Evolution modern humans in Africa	200,000 years ago	*NYTimes*, Science
Agricultural seeds + sheep brought from Nibiru to Earth	108,000 years ago?	Tellinger

Events	Dates (yr)	Sources
The main exodus of *H. s. sapiens* from Africa	80,000—50,000 years ago.	*NYTimes*, Science
Europe: Wurm glaciation	50,000—17,000 yrs ago	Hancock
Cave paintings in Europe	35,000 years ago	Tellinger
Neanderthal man extinct	30,000 years ago	Tellinger
Sudden meltdown and earth crust displacement	14,500–12,500 BC	Hapgood
Floods in the Nile Valley	13,000–10,500 BC	Hancock
Nibiru last inside solar system	11,000, 7,400, and 3,800 BC	Sitchin
The Great Flood	11,000 years ago	Evans
Pyramid building predated (by water runnels on the Sphinx—when climate was rainy)	2,500 BC to ≈ 10,500 BC	Hancock
Oldest pottery; oldest domestic crops	10,000 years ago	Tellinger
Mammoths extinct; 'Antarctica' moved to S. Pole	8,000 years ago	Tellinger
The start of city rebuilding	7,400 years ago	Evans
Period of deglaciation	7,000 years duration	Hancock
Oldest human city	3,500–3,700 years ago	Tellinger
End of Sumerian civilization	End of 3rd millennium BC	Sitchin
Nuclear conflict adjacent to spaceport in Sinai/Gaza.	2,024 BC	Evans
Orbital timing: planet Nibiru	Every 3,600 Earth yrs	Sitchin
Sumerian clay tablets	3,000–5,000 years old: many recopied from older originals	

A Brief Summary of the Argument

This short book reflects my approach as a scientist thinking about the origin of intelligent life on our planet and the evolution of human intelligence. I am not going to assert categorically that the human evolution to intelligence was entirely a result of the actions of extra-terrestrials of advanced intelligence, but the hypothesis that our development to an intelligent urban species was in part due to this was found compatible with a variety of other information sources.

Although this analysis of published material cannot be regarded as definitive, I have investigated the hypothesis that we were helped in our evolution to civilization by intelligent entities who long preceded us in our galaxy. The first action by a very early extraterrestrial species, through directed panspermia, seems to have been to inoculate DNA or living cells onto our planet once its surface had cooled adequately. This allowed syntropy to speed up evolution faster than would have been possible if the natural synthesis of DNA was by unaided Darwinian processes, which has been estimated to require a period longer than the duration of life on earth. It is supposed that different intelligent forms evolved successively over the long history of our planet, interrupted by the periodic climatic catastrophes typical of our planet.

Evidence suggests that our further evolution beyond the hunter-gatherer stage was aided by galactic influences on our DNA and by speeding up access to the technology for off-planet space travel. We can deduce that the eventual purpose of this technology is to ensure that despite the periodic catastrophes planet Earth is subject to, the extinction of intelligent planetary populations could be avoided, allowing mankind to join other intelligent species in a cosmic multispecies civilization.

Planet Earth has had a long history as an incubator for intelligent life forms, and these are obliged to rapidly develop to intellectual maturity between successive planetary catastrophes. There seems some evidence that reptilian forms of intelligence may have existed on earth before the Age of the dinosaurs was brought to an end by asteroid bombardment. A range of disasters includes periodic Ice Ages, earthquakes, volcanic eruptions, and extreme atmospheric temperature fluctuations are all typical and regular occurrences on our planet. It is essential to avoid adding to environmental catastrophies by destructive warfare or pollution of the atmosphere and water bodies, and by adding toxic radioactive elements to the global ecosytem.

Following from Zacharia Sitchin's conclusions on the Anunnaki, it seems very probable that intelligent life forms who had already made the transition off their own planet were involved in our transition to urban civilizations. The Anunnaki were conventionally viewed by the first translators of cuneiform writing as the 'divinities' of the ancient Sumerians. From a modern perspective however, there seems adequate information to support Sitchin's idea that they were an extraterrestrial species from a planet called Nibiru, which is hypothesized to be rotating around a dwarf star with an orbit intersecting the solar system only at long intervals. Following the last Ice Age, there seems evidence for the Anunnaki influencing the very rapid growth of urban civilizations from hunter-gatherer cultures to large cities in Mesopotamia and Egypt, with trade and irrigated agriculture and advanced knowledge of the long-term cycles of the earth—all without any obvious human evolutionary stages.

The subsequent extension of Megalithic civilizations worldwide saw the birth of common supporting technologies that are not as yet understood by contemporary science. A genetic intervention by the Anunnaki to convert pre-humans to effective workers is addressed in Sumerian accounts, and recent studies of human genotypes do not contradict the idea that the pre-hominids encountered by the Anunnaki were genetically modified to become useful workers, such that their descendants eventually gave rise to modern human beings. Nonetheless, the existence of an extraterrestrial 'tutor' for our species has not been readily accepted by some authorities.

The recent discovery of an extensive group of enclosures and associated terraces for food production in southern Africa dating back some 200,000 years or more, suggests that these may have been constructed to accommodate and feed human slave labour working in the ancient goldmines in the vicinity. The possibility of a subsequent northward migration of modified humans and Anunnaki to Sumer and ancient Egypt is postulated.

Biblical texts from Genesis and the book of Enoch parallel earlier Sumerian accounts from just after the Great Flood, such as the stories of the Garden of Eden, Noah's Ark, The Tower of Babel, and the descent of angels to wed the daughters of men. These were interpreted as actual events in Sumerian accounts. The last-mentioned cross-breeding of Anunnaki and modified proto-humans also described in Sumerian stories, paralleled accounts from the Book of Enoch, and led to a giant hybrid species whose remains are found worldwide but which has since disappeared. These hybrids were characterized by an elongated skull, leading to the term 'cone-heads', but they could possibly have been the Nephilim giants referred to in ancient scriptures. The cone head hybrids were associated with the megalithic edifices built in different parts of the world, and it is supposed that the unknown technologies used in their construction were derived from Anunnaki knowhow.

Although human tribes imitated the Anunnaki skull configuration by compressing the heads of their infants between boards, this was probably to gain favour with their megalithic masters and led to a much smaller brain capacity and different skull sutures. Preliminary DNA analysis of cone-head skulls from Peru suggests that their mitochondria contain extra-terrestrial as well as human components. Skulls of giant cone-heads have been discovered in ancient burials around the world, but as for other unexpected or unexplained developments, public information has often been suppressed on their involvement in our early history. Despite their very much longer lifespans than humans, the hybrids between the Anunnaki and humans may have been low in fertility, and eventually were replaced by reproductively more fecund human populations after the last Ice Age.

Several sources of information confirm that much earlier civilizations of extraterrestrials were, and possibly still are, present on Mars and the moon; in the second case, it is supposed that monitoring evolution on Earth was one of their objectives.

Extensive warfare between social or hereditary groups of cone-heads aided by their human followers, eventually used the nuclear weapons brought from Nibiru. As evidence for an ancient period of armed conflict, vitrified patches of ground have been found in different parts of the world caused by nuclear explosions. Evidence also exists that points to nuclear warfare as responsible for the extinction of intelligent species and their civilizations on other planets. This conclusion from pre-history should give the powers that be just cause to eliminate these deadly weapons from their armouries.

A Dedication and an Afterword

The author expresses his love and gratitude to his wife, Franca, whose life came to an end as this book approached publication. The above is Franca when she was 13 years old, and this ancient sketch expresses her positive attitude to others which persisted throughout life. By its subject matter this book may have raised negative emotions from the material treated, and expressing positive emotions to others seems essential at this point.

If we can avoid reinforcing the habits installed in us by the Anunnaki when we encounter human or extraterrestrial visitors, a simple remedy would be to greet them with friendship and a smile, as befits members of our extended family of intelligent beings.

Printed in the United States
By Bookmasters